Sacred Narcotic Plants

of the NEW WORLD INDIANS

An anthology of texts from the sixteenth century to date

Compiled by Hedwig Schleiffer

Introductory words by Richard Evans Schultes
Professor of Biology
Director, Botanical Museum Harvard University

Hafner Press
A DIVISION OF MACMILLAN PUBLISHING CO., INC., NEW YORK
COLLIER MACMILLAN PUBLISHERS, LONDON

Hafner Press
A Division of Macmillan Publishing Co., Inc.
866 Third Avenue, New York, N.Y. 10022
Collier-Macmillan Canada Ltd.

Library of Congress Catalog Card Number: 73-91363
ISBN: 0-02-851780-6

Printed in the United States of America

Contents

Acknowledgments

It is a pleasure to have this opportunity to thank the administration of Harvard University Library, in particular Mr. Douglas W. Bryant, Director, and Mr. Theodore Alevizos, Associate University Librarian, and the staff of the Reading Room of Houghton Library for all the facilities they have made available to me.

Above all my thanks are due to Dr. Richard E. Schultes for the assistance, encouragement and time he always has generously given to me. I am further indebted for the courtesy and help accorded me by the staff of the Economic Botany Section of the Botanical Museum, Harvard University.

Introductory Words

There is little that can put a custom more sharply in relief than the literature arising around its history and place in a culture. Yet this simple fact seems frequently to go unregarded.

The significance of the literature that has grown up around the narcotics has not yet been fully recognized. This literature is extensive and diverse. Somehow its wealth and expressiveness have escaped even many of those readers most closely touched by the use of these substances capable of such unearthly effects upon man's mind and body.

Narcotic plants have been part of men's living for many millenia. The hallucinogens—by any yardstick the most astounding of the narcotic agents—have even become sacred, and have assumed exalted roles in the magic, medicine, religion, mysticism and mythology of many primitive societies. The hallucinogens have recently captured the imagination of man in modern, sophisticated cultures: sometimes because of their promise in the study of mechanisms of hallucinations; sometimes because of their misuse or abuse for trivial hedonistic purposes in certain "subcultures" of western society.

Literature about the narcotics—especially the hallucinogens—boasts a long history. Its antiquity in the Old World, where writing goes back many millenia, is much greater than in the New. Written records concerning the narcotics in the Americas, notwithstanding their younger stature, seem to indicate a deeper, more penetrating, more intimate part in native life. This apparent disparity between the two hemispheres may be due in part to the much greater diversity of narcotic plants employed in the New World, but it may also reflect a fundamentally more significant role in the life and being of the American aborigine.

For this anthology, Dr. Schleiffer has thoroughly searched the literature on the American narcotics and has culled from it some of the most meaningful gems. Her choice of selections is historically and culturally well balanced. She has gone back to the earliest written reports by the Europeans who first encountered the use of some of these extraordinary plants. She brings the reader up through the turbulence of the cultural disruptions that engulfed native Americans following their contacts with European civilization. She leads us into a consideration of various aspects of the recent birth of interest in hallucinogens in our own society.

It has not been an easy task to assemble this anthology. The literature, especially the modern, is extensive and comprises much that is good and much that has a casual, even superficial and trivial, nature. Not only has it been in modern times and sophisticated cultures that the utilization of narcotics has been frowned or lauded upon. It seems hard to find a period when these plants have not been subject to the most extravagant praises and to the most damning condemnations. Realizing this, Dr. Schleiffer has striven scrupulously to take quotations representative of the diverse points of view. She has, I believe, been eminently successful in her choices.

Why an anthology on sacred narcotic plants of the New World? Perhaps there are several good reasons for such a collection. First: these writings are highly interesting in and of themselves. Second: they represent an often neglected aspect of cultural history. Third: they afford us a clearer insight into the striking and brilliant mind of the native Americans. Fourth: it is often—nay usually—impossible to find many of these extraordinary quotations, even when rather extensive library facilities are available. And fifth: if, indeed, we are experiencing nascent or growing drug threats in our own societies, an understanding of the special role of narcotics in less advanced cultures might conceivably help us immeasurably in coping with the problems ahead.

As an investigator who has devoted some forty years to an ethnobotanical study of New World narcotic plants, both in the field and in the laboratory, I commend Dr. Schleiffer on the great service she has done in gathering together this anthology and in presenting it to today's public—an audience that seems to be desirous of increased enlightenment in so far as narcotic plants, their place, their dangers, their innocuousness, their future in our culture are concerned.

RES

The Narcotic Complex

LOUIS LEWIN

The passionate desire which consciously or unconsciously leads man to flee from the monotony of everyday life, to allow his soul to lead a purely internal life even if it be only for a few short moments, has made him instinctively discover strange substances. He has done so, even where nature has been most niggardly in producing them and where the products seem very far from possessing the property which would enable him to satisfy this desire.

From Louis Lewin (1850-1929) *Phantastica. Narcotic and Stimulating Drugs, Their Use and Abuse.* Forword by Bo Holmstedt. Routledge & Kegan Paul Ltd. London (1931) pp. 123-124. (Reprinted by permission)

M[ORDECAI] C[UBITT] COOKE

Mind and body alike suffer from the want of sleep, the spirit is broken, and the fire of the ardent imagination quenched. Who can wonder that when disease or pain has racked and tortured the frame, and prevented a subsidence into a state so natural and necessary to man, he should have resorted to the aid of drugs and potions, whereby to lull his pains, and dispel the care which has banished repose, and woo back again—

'the certain knot of peace,
The baiting place of wit, the balm of woo;
The poor man's wealth, the prisoner's release,
Th' indifferent judge between the high and low.'

From M[ordecai] C[ubitt] Cooke (1825-1914) *The Seven Sisters of Sleep. Popular History of the Seven Prevailing Narcotics of the World.* James Blackwood. London. (1860) p. 7

LOUIS LEWIN

If human consciousness is the most wonderful thing on

earth, the attempt to fathom the depth of the psycho-physiological action of narcotic and stimulating drugs makes this wonder seem greater still, for with their help man is enabled to transfer the emotions of everyday life, as well as his will and intellect, to unknown regions; he is enabled to attain degrees of emotional intensity and duration which are otherwise unknown to the brain. Such effects are brought about by chemical substances. The most powerful of these are products of the vegetable kingdom, into whose silent growth and creative abundance man has not yet fully penetrated . . .

Miracles like these are performed throughout the world by these strange substances wherever men are in possession of any one of them. The savage in the jungle beneath a sheltering roof of leaves and the native of the storm-swept island secured through these drugs a greater intensity of life. The solitary dweller in a distant mountain cavern can with their aid relieve the dull monotony of his cramped existence. Various are the motives which induce civilized men to such a transient sensation of pleasure. The potent influence of these substances leads us on the one hand into the darkest depths of human passion, ending in mental instability, physical misery, and degeneration, and on the other to hours of ecstasy and happiness or a tranquil and meditative state of mind.

. . . A large field of activity and research is offered to psychology which hitherto was open only to a few scientists. . . .To the ethnologist numerous problems, promising much new enlightenment, not the least those relating to the study of comparative theology, present themselves with regard to the extension and cause of the use of these drugs.

From Louis Lewin (1850-1929) *Phantastica. Narcotic and Stimulating Drugs, Their Use and Abuse.* Foreword by Bo Holmstedt. Routledge & Kegan Paul Ltd. London (1931) pp. XV-XVI. (Reprinted by permission)

RICHARD EVANS SCHULTES

The Plant Kingdom has always been man's main source of the necessities and the amenities of life. Since the prime necessity is and always has been food, man must have eaten experimentally every conceivable kind of plant material in times of hunger. Some of the plants which he put into his stomach were outright poisons with which he was physically unable to experiment more than the first time. There were others, however, which, although they did act in the same ways like poisons, induced physical and mental states not at all unpleasant and oftentimes of startling unreality. Man had then become familiar with narcotic plants.

As his sophistication increased, man found it necessary to try to explain these extraordinary powers of some of the plants in his environment. In all primitive cultures, this explanation invariably ascribed to the plants some particular divinity or spirit which, in many instances, was thought to be efficacious as an intermediary between man's world of humdrum reality and the supernatural or spirit realm.

The use of narcotics is always in some way connected with escape from reality. From their most primitive uses to their applications in modern medicine or their abuse in modern society this is true. All narcotic plants have also, sometime in their history, been linked to religion or magic. This is so even of tobacco, coca and opium, which have suffered secularization and are now used hedonistically. Some narcotics—peyote is an example—conserved even today this religious basis in their use. And it is interesting to note parenthetically that when problems do arise from the employment of narcotics, they arise after the narcotics have passed from ceremonial to purely hedonic or recreational use.

From Richard Evans Schultes "Hallucinogenic Plants of the New World." *The Harvard Review* (Harvard University) Cambridge Mass. 1(1963) pp. 4, 18. (Reprinted by permission)

PETER T. FURST

Perhaps the discovery that certain substances found in nature help man to move beyond his everyday experiences to "Otherworlds" and the institutionalizing of these personal ecstatic experiences into an ideological and ritual framework accepted by the group as a whole (i.e. religion, but not necessarily an organized cult) goes back almost to the beginnings of human culture. Shamanism as a universal *Ur*-religion, which eventually gave rise to various cults, including the great world religions, must reach deep into the Paleolithic—at least as far back as the oldest known deliberate interments of the dead by the Neanderthal man, *ca.* 100,000 years ago, and perhaps hundreds of thousands of years earlier. The striking similarities between the basic premises and motifs of shamanism the world over suggest great antiquity as well as the universality of the creative unconscious of the human psyche.

Wherever shamanism is still encountered today, whether in Asia, Australia, Africa, or North and South America, the shaman functions fundamentally in much the same way and with similar techniques—as guardian of the psychic and ecological equilibrium of his group and its members, as intermediary between the seen and unseen worlds, as master of spirits, as supernatural curer, etc. He seeks to control the weather and to ensure the benevolence of ancestor spirits and deities. Where hunting continues to have importance (ideologically or economically), he ensures, by means of his special powers and his unique psychological capacity, to transcend the human condition and pass freely back and forth through the different cosmological planes (as these are conceived in the particular world view of his group), the renewal of game animals and, indeed, of all nature. . . .

. . . Be it mushrooms, yajé, peyote, tobacco, . . . the psychoactive plant in the traditional culture transports the user to a "land beyond", whose geography he already knows because he has heard it described innumerable times before; i.e., the values of the parental generation. He already has the answers, so that for him the psychedelic experience is in part, at least, a quest for their confirmation. It is the means to a known end, not, as for so many youthful devotees of the Western cult of psychedelia, an end in itself.

Despite these fundamental differences, the role of hallucinogens in societies other than ours has considerable relevance to our own time and place. If nothing else, understanding the function and the social and physiological effects of the abundant sacred hallucinogenic pharmacopoeia among non-Western peoples may, on the one hand, help lower the level of public hysteria about real or fancied dangers in the use of hallucinogens and, on the other, make drug users themselves more aware of spurious versus genuine culture. Ideally, it may also help in the formulation of more realistic and intelligent drug legislation. The fact is that we still know far too little of this important area of the study of man. More than four centuries after Sahagún described mushroom and peyote ceremonies in Mexico, and after a century of anthropological field work among so-called primitive peoples, many of whom had a wide knowledge of psychoactive plants in their environment, we are only just beginning to consider the real cultural, psychological, historical, and pharmacological significance of these potent substances. But while the specialized scientific literature on the chemistry and physiological effects of hallucinogens has grown by leaps and bounds, with the exception of the North American Indian peyote cult (properly the Native American Church), relatively little attention has been paid to those societies that long ago institutionalized the psychedelic phenomenon—with ritual rather than legal controls—and that have employed hallucinogens for centuries without the dire social and individual consequences so often ascribed to sustained use of these narcotics. There is an obvious need, then, to illuminate the role of the cultural variable in the use of hallucinogens, the important and often decisive parts these drugs have played and continue to play in a variety of sociocultural contexts (especially in religious belief and ritual and the preservation of cultural independence and integrity), and the part they might have played historically in the remote origins of religion itself.

From Peter T. Furst Introduction to *Flesh of the Gods.* (c) 1972 by Praeger Publishers, Inc., New York. pp. VIII, IX, XIII, XIV. (Reprinted by permission)

RICHARD EVANS SCHULTES

There is an urgency for research into hallucinogenic plants amongst aboriginal peoples. Civilization is on an ever more accelerated march, penetrating most regions of the world still held by primitive societies. The consequent divorcement of aboriginal peoples from dependence upon their vegetal environment for the necessities and amenities of life has been set in motion; nothing will now check it. The resulting disintegration of knowledge of plants and their properties is frightening. Our challenge must be to salvage native botanical lore—especially that relating to folk medicine in its broadest sense—before it becomes forever entombed with the cultures that gave it birth.

From Richard Evans Schultes "The New World Indians and their Hallucinogenic Plants." *Morris Arboretum Bulletin.* Philadelphia, Pa. 21 (1970) pp. 3-4. (Reprinted by permission)

NORMAN R. FARNSWORTH

. . . Most of our definitive chemical knowledge of hallucinogenic plants has evolved only during the past decade or so. Three important factors have been responsible for this. First, valid botanical authentication of many hallucinogenic plants has been accomplished only during this period. Prior to this, chemical reports on these plants were, to say the least, chaotic. Second, the isolation and identification of active principles has been enhanced by the introduction of modern research techniques which require only small amounts of plant material. Finally, there has been an acute need for the active principles present in hallucinogenic plants, either as potentially useful drugs in the treatment of mental disease, or as new tools for the pharmacologist in his attempt to shed some light on the biochemical causes of mental illness. These needs have accelerated the research, in which workers from many disciplines have participated: botanists, ethno-botanists, ethnologists, chemists, pharmacognosists, pharmacologists, psychologists, and others.

From Norman R. Farnsworth "Hallucinogenic Plants." *Science.* Washington, etc. 162 (1968) p. 1086. (Reprinted by permission)

RICHARD EVANS SCHULTES

Man in primitive societies the world around has found the most ingenious ways of administering narcotics. Intoxicating plants, or products from them, have been chewed in crude form or variously elaborated and consumed. They have been drunk as decoctions or infusions. A few have been prepared in the form of thick syrups or pastes that are licked or smeared on the tongue or gums. Some have been smoked directly, as in pipes, cigars or cigarettes, or the fumes of them have been inhaled in sundry ways. There are those that have been applied to the skin or membranes in the form of ointments or unguents. Several are known to have been taken as an enema. Snuffing has been the preferred method of using a number of these agents.

From Richard Evans Schultes "The Botanical Origin of South American Snuffs." In (D. H. Efron, ed.) *Ethnopharmacologic Search for Psychoactive Drugs,* p. 291 (Public Health Service Publication #1645) U.S. Government Printing Office, Washington, D.C. (1967) (Reprinted by permission)

ALBERT HOFMANN

Unlike the true narcotics of the opium and heroin class, etc., whose injurious effects develop in a chronic fashion, the hallucinogens do not give rise to addiction. Much is made of this by the proponents of the release of hallucinogens for general use to support the innocuous nature of these substances. However, the hallucinogens are not less dangerous than the drugs of addiction; they are dangerous in a different way. The great danger associated with the use of hallucinogens without medical supervision is the possible occurrence of a 'bad trip', i.e., the development of severe confusional and anxiety states which may lead to suicide, or in psychically labile or youthful persons whose character is still not mature, permanent psychic traumata. The power of the Mexican hallucinogens to completely transform perceptivity, and the dangers inherent in such a process, readily explain why the primitive peoples imposed a taboo on these plants and fungi. For these peoples they were sacred and they were reserved for the medicine men for use in religious, ceremonial situations. Since in our society taboos no longer exist and the general trend is to throw aside the last inhibitions, the authorities have no alternative but to impose strict controls, restricting the use of hallucinogens to scientific research and medical applications. Legal measures are not an ideal solution to the drug problem, but in the present situation and in our present-day society they are the only choice for the time being.

From Albert Hofmann "Teonanacatl and Ololiuqui, Two Ancient Magic Drugs of Mexico." *Bulletin on Narcotics.* (United Nations. Division of Narcotic Drugs, Department of Economic and Social Affairs.) New York. 23 (1971) p. 12. (Reprinted by permission)

WESTON LA BARRE

It strikes us as strange that the 'doctor' takes the 'medicine' in the whole area of the American narcotic complex, rather than the sick patient. But this is entirely logical in native terms, since it is the shaman who needs the supernatural 'power' to effect a cure, i.e., to diagnose the human or physical cause (often a crystal, a feather, a claw to be sucked out), to contest a rival's malevolent magic causing the illness, to prognosticate, for clairvoyance, to control the weather, etc. . . .

For the American Indian, the presence of any psychotropic effect in a plant was plain evidence of its containment of supernatural 'medicine' or spirit-making 'power'. One introjected the power exactly as he ate food. This principle was true of even so mildly psychedelic a drug as tobacco. . . .

From Weston La Barre "Old and New World Narcotics: A Statistical Question and an Ethnological Reply." *Economic Botany.* Lancaster, Pa. 24 (1970) pp. 76-77. (Reprinted by permission)

JUAN DE CARDENAS

Last Chapter in which is fully explained whether plants can have the power to bewitch and what bewitching is.

. . . There remains to be mentioned what has been learned in the Indias about the *Peyote* [*Lophophora Williamsii*], the *Poyomate* [*Salvia divinorum*], the *holofquen* [*Ololiuhqui-Rivea corymbosa*], and the *Piciete* [*Nicotiana rustica*].

Many, chiefly Indians, Negroes, or stupid and ignorant people, affirm that the aforesaid herbs, when eaten, can summon the demon to appear. The demon then informs them about things to happen in the future.

Now it is opportune to establish whether any herb or root has the innate strength and power to compel the demon to appear and whether through its innate power we may divine the future. To this I may reply that part of this power is in the plant, and only part of it should be attributed to the demon. Furthermore, I wish to state that if any of the above herbs, or any other herb with similar power, were eaten or used, it would by its nature cause three things to happen in the human body. All the rest is illusion and work of the demon.

These herbs produce the following effects: first, they have an innate strength and heat and consist of subtle, strong and hot parts; when these enter the stomach, the natural (body) heat begins to alter and heat them more. When thus heated, they rise in the form of vapors to the brain and all parts of the body. When these subtlest, strongest and hottest parts of the herb rise to the brain, and to all parts, passages and pores in the body, they begin to foment, thus disturbing and confusing the animal vigor of the body. In this way they deprive a man of his senses, an effect similar to that of wine and *piciete* (tobacco). Finally, any herb, nay any strong and vaporous drink or food, may by its nature produce the same effect. This is the first effect which the herb or root produces by its innate power.

The second effect is to induce in its users a heavy and tormenting sleep. This results from the same thick and vaporous humors, which, although first subtle, become stronger because of the coldness of the brain. Thus they will induce not a peaceful, light and refreshing sleep like that resulting from the gentle and humid vapors of food, but a horrible and frightful sleep.

The third effect of the aforesaid herbs, or of their distressing humors, is to disturb and confuse the images in the mind within the brain. If these images become confused, they come to represent in man's imagination not joyful visions and things which would restore his imaginative faculty, but rather horrid and frightful pictures, or forms of monsters, bulls, tigers, lions, or phantoms—in short of horrible and distressing things. It could not be otherwise. If such a sleep and visions are brought about by black, strong and heavy humors, it is understandable that these visions cannot be of fine, beautiful, attractive or pleasant things, but rather of beasts or horrible things, such as the demon, who in a vision must appear as a horrible monster.

By its innate power, any of the above listed herbs may produce all three of these effects. If a person takes such an herb, it has the power of depriving him of his senses and inducing in him a horrible sleep disturbed by visions of horrible things. Now, there must be a motive or reason for an herb or root to do what it cannot ordinarily do: namely, a pact or communication with the demon. Here I wish to say this: first, the sick person summons the demon to appear—the herb cannot do this, indeed it would be very wrong to say that an herb by its innate power could compel the demon to appear. Second, it would be a notable error to say that by its innate power an herb knows of future things and of secret things of the past—things which only the demon can communicate. I understand that in ancient times, such lies came from the mouths of the oracles of false gods. However I consider it wrong and a lie to consider that a plant by its innate power can do any such thing. What I believe in this regard is that the demon, having hoodwinked and deceived some unfortunate person, will suggest that he uses any of these herbs—and not that this herb by its innate power can compel the demon to appear to a person at his request. I rather believe that a person by taking this herb gets intoxicated and loses his senses. Having become mad, he will lose his fear of so ugly and horrible a thing as the demon must be. Thus having lost his senses and become awe-struck, he sees the demon, who appears in order to communicate with and deceive him, talking and replying to what he asks. . . .

From Doctor Juan de Cardenas (1563-1609) *Primera Parte de los Problemas y Secretos Maravillosos de las Indias.* (Translation: First part of the problems and marvelous secrets of the Indias.) Pedro Ocharte. México. (1591) fol. pp. 243v-245v. (Book III, last chapter).

Agaricaceae or Mushroom Family

The Divine Mushroom

R. GORDON WASSON

The little mushroom comes of itself, no one knows whence, like the wind that comes we know not whence nor why.

From R. Gordon Wasson (1898-) "The Hallucinogenic Fungi of Mexico: An Inquiry into the Origin of the Religious Idea among primitive Peoples". ('Mexican Muleteer'). *Botanical Museum Leaflets.* Harvard University, Cambridge, Mass. 19 (1961) p. 147. (Reprinted by persmission)

R. GORDON WASSON

As man emerged from his brutish past, thousands of years ago, there was a stage in the evolution of his awareness when the discovery of a mushroom (or was it a higher plant?) with miraculous properties was a revelation to him, a veritable detonator to his soul, arousing in him sentiments of awe and reverence, and gentleness and love, to the highest pitch of which mankind is capable, all those sentiments and virtues that mankind has ever since regarded as the highest attribute of his kind. It made him see what this perishing mortal eye cannot see. . . . What today is resolved into a mere drug, a tryptamine or lysergic acid derivative, was for him a prodigious miracle, inspiring in him poetry and philosophy and religion. Perhaps with all our modern knowledge we do not need the divine mushrooms any more. Or do we need them more than ever? Some are shocked that the key even to religion might be reduced to a mere drug. On the other hand, the drug is as mysterious as it ever was: 'like the wind it cometh we know not whence, nor why'. Out of a mere drug comes the ineffable, comes ecstasy. It is not the only instance in the history of mankind where the lowly has given birth to the divine.

From R. Gordon Wasson (1898-) "The Hallucinogenic Fungi of Mexico: An Inquiry into the Origins of the Religious Idea among Primitive Peoples." *Botanical Museum Leaflets.* Harvard University, Cambridge, Mass. 19 (1961) p. 57 (Reprinted by permission)

For more than four centuries the Indians have kept the divine mushroom close to their hearts, sheltered from desecration by white men, a precious secret. We know that today there are many *curanderos* who carry on the cult, each according to his lights, some of them consummate artists, performing the ancient liturgy in remote huts before miniscule congregations. With the passing years, they will die off, and, as the country opens up, the cult is destined to disappear.

From R. Gordon Wasson (1898-) "The Hallucinogenic Fungi of Mexico: An Inquiry into the Origins of the Religious Idea among Primitive Peoples." *Botanical Museum Leaflets.* Harvard University, Cambridge, Mass. 19 (1961) p. 147. (Reprinted by permission)

. . . There exist certain species of wild mushrooms that contain a most potent and mysterious drug—mushrooms which, if you eat them, cause you to see visions, breathtakingly vivid visions in color and motion, visions of almost anything you can imagine except the scenes of your everyday life. . . .

. . . From earliest times they have been worshipped by certain primitive peoples scattered from Mexico to Borneo and Siberia, and we think formerly in Europe too.

The visions that you see when you eat the mushrooms are staggering in their subjective impact. They are no shadowy, uncertain sights. When you see them you are moved to exclaim, if only to yourself, that never in your normal state have you seen things so clearly, so truly. At last you are seeing clearly, not as through a glass darkly. If in that span of ecstatic vision you think of Plato, you will say to yourself that you are seeing the Archetypes, the very Ideas of Plato. You do not lose possession of yourself. You are in a pseudo-schizophrenic state, and you can take notes, and observe yourself, and afterwards you remember everything in detail. Possibly the hallucinations are aural as well as visual. Just as you are seeing things in the mind's eye, so you may hear in the mind's ear, the music of the spheres accompanying the Ideas of Plato. The emotional effects of the drug are equally interesting. You experience ecstasy. In my case, for the first time the meaning of 'ecstasy' came over me, no longer as an intellectualized definition, but subjectively. Indeed, you are spellbound by awe, by feelings of wonder and reverence, by an overflowing sense of empathy, of *caritas* towards those who are sharing the mushrooms and the experience with you.

The primitive peoples who worship these mushrooms consider that they open the gates to another plane of existence, to the past and future, to Heaven and God, who then answers truly all grave questions put to him. . . .

On this subject I speak to you not by hearsay but as a witness. On the night of June 29-30, 1955, my friend and photographer Allan Richardson and I ate these mushrooms in the course of a mushroomic communion or agape in a remote circle of Indians in Mexico. So far as the records indicate, we are the first white men of the modern world to have had this experience. We underwent the visual, aural, and emotional effects of the mushrooms. . . .

From R. Gordon Wasson (1898-) "The Divine Mushroom: Primitive Religion and Hallucinatory Agents." *Proceedings* of the American Philosophical Society. Philadelphia, Pa. 102 (1958) pp. 221-23. (Reprinted by permission)

Teonanácatl or Nanácatl

BERNARDINO DE SAHAGUN

"On Certain Intoxicating Plants."

. . . There are some small mushrooms native to this country, called teonanácatl by the Indians. They grow under moss in the fields or in the high and cold regions. They are round and have a rather high, slender and round stem. When eaten, they have a bad taste, hurt the throat and intoxicate. They are medicinal against fever and gout; one must eat two or three but not more. Those who eat them see visions and feel a faintness in their hearts. Those who eat many of these mushrooms are provoked to lust, even if they ate few. (Book XI, ch. vii).

. . . The [Chichimecas] also were well acquainted with herbs and roots and knew their properties and powers. They discovered, and were the first to use the root which they call *péyotl;* they ate and drunk this instead of wine and in the same manner they used the *nanácatl* which are wicked mushrooms, also intoxicating like wine. Having drunk and eaten these, they gathered in a plain where, for their pleasure, they danced day and night. This they did on the first day, for on the next day they all cried very much saying that they cleaned and washed their eyes and faces with their tears. (Book X, ch. xxix, paragraph 2.)

. . . At a banquet the first thing the Indians ate, was a black mushroom which they call *nanácatl.* These mushrooms caused them to become intoxicated, to see visions and also to be provoked to lust. They ate the mushrooms before dawn when they also drank cacao. They ate the mushrooms with honey and when they began to feel excited due to the effect of the mushrooms, the Indians started dancing, while some were singing and others weeping. Thus was the intoxication produced by the mushrooms. Some Indians who did not care to sing, sat down in their rooms, remaining there as if to think. Others, however, saw in a vision that they died and thus cried; others saw themselves being eaten by a wild beast; others imagined that they were capturing prisoners of war; others that they were rich or that they possessed many slaves; others that they committed adultery and had their heads crushed for this offence; others that they had stolen some articles for which they had to be killed, and many other visions. When this mushroom intoxication had passed, the Indians talked over amongst themselves the visions they had seen. (Book IX, ch. viii)

From Fray Bernardino de Sahagún (1499-1590) *Historia General de las Cosas de Nueva España* (Translation: General History of Things of New Spain (Mexico) 5 volumes. Pedro Robredo. Mexico, D.F., 1938 (or any other edition).

HERNANDO ALVARADO TEZOZOMOC

"How the Great Sacrifice was Celebrated in Honor of 'Huitzilopochtli' and in Honor of Emperor Moctezuma and the Mexican Council and how with Courtesy were Bidden farewell the Foreign Lords who were much Satisfied with Having Seen The Great Cruelty which they had never Seen before."

. . . The foreigners were given to eat mushrooms grown in the forests and with these mushrooms they got drunk and entered the dance. . . .

From D. Hernando Alvarado Tezozomoc (Fl. 16th Century) *Crónica Mexicana Escrita . . . hacia el Año de MDXCVIII. . . . Precedida del Codicil RAMIREZ,* Manuscrito del Siglo XVI intitulado: *Relación del Origin de los Indios que Habitan esta Nueva España segun sus Historias.* (Translation: Mexican Chronicle written down to the year 1598. . . . Preceded by the RAMIREZ Codex, a manuscript of the 16th century, entitled: Account of the origin of the Indians inhabiting New Spain (Mexico), according to their history) I. Paz, México. (1878) ch. lxxxvii of the "Relation."

FRANCISCO HERNANDEZ

"Of the Nanácatl or a Kind of Mushroom"

In Nueva España [Mexico] grow many and so different mushrooms that it would be too lengthy and tiresome to describe and depict every one of them. Therefore I will limit myself and give a detailed description only of some and leave the rest to another occasion. . . . Therefore we shall say that certain mushrooms native to this country are deadly poisons. These are called *citlalnanacame;* there are others, called *teihuinti* which, when eaten, do not cause death, but sometimes produce a certain insanity manifesting itself in an immoderate hilarity. These are tawny, acrid and have a strong but not unpleasant odor. There are other mushrooms which, without causing laughter, produce all kinds of visions, such as wars, and the likeness of demons. Yet others, with no less wicked and horrid [properties] are desired by the headmen, especially for their feasts and banquets, and are gathered for a high price and with great care. These mushrooms are brown and somewhat sharp in taste.

From Dr. Francisco Hernandez (1515-1587), Personal Physician to the Emperor. *Opera. . .* (Translation: Works) 3 volumes. Printed by Ibarra's Heirs. Madrid. (1790) Book IX, ch. xcv. (Vol. 2, p. 357)

MOTOLINIA TORIBIO DE BENAVENTE

[In addition to getting drunk with their native wine] they had another way of becoming intoxicated which made them even more cruel. This the Indians did with some fungi or small mushrooms native both to this country and to Castile. But those in this country are of such nature, that when eaten raw and being bitter, they eat and drink them together with some honey of the bees and shortly afterwards they see a thousand visions, especially of snakes. But being completely out of their mind, the Indians imagined to have their legs and bodies covered with worms which were eating them alive. Thus half mad they rushed from their house, hoping that someone would kill them; while being beastly drunk and experiencing this curse, it happened sometimes that they hung themselves and also became more cruel toward others. In their native language these fungi are called Teonanácatl, which means "flesh of the gods", flesh of the demon whom they worshipped. Thus their cruel gods communicated with them through this bitter food.

From Fray Toribio Motolinia (d.· 1568) *Historia de los Indios de la Nueva España.* (Translation: History of the Indians of New Spain (Mexico).) Heirs of J. Gili, (1914) First Treatise, ch. ii.

Sitho, 'nti-sitho

VALENTINA PAVLOVNA
and R. GORDON WASSON

[A letter addressed to Mr. Wasson by Miss E. V. Pike, an American linguist, who had been living among the Mazatec Indians for many months each year since 1936.]

Huantla de Jiménez, Oaxaco, Mexico
March 9, 1953

Dear Mr. Wasson:

I'm glad to tell you whatever I can about the Mazatec mushroom, . . . Mazatecs seldom talk about the mushroom to outsiders, but belief in it is widespread. A twenty year old boy told me, I know that outsiders don't use the mushroom, but Jesus gave it to us because we are poor people and can't afford a doctor and expensive medicine.'

Sometimes they refer to it as 'the blood of Christ' because supposedly it grows only where a drop of Christ's blood has fallen. They say that the land in this region is 'living' because it will produce the mushroom, whereas the hot dry country where the mushroom will not grow is called 'dead'.

They say that it helps 'good people' but if someone who is bad eats it, 'it kills him or makes him crazy'. When they speak of 'badness' they mean 'ceremonially unclean'. (A murderer if he is ceremonially clean can eat the mushroom with no ill effects.) A person is considered safe if he refrains from intercourse five days before and after eating the mushroom. A shoemaker in our part of town went crazy about five years ago. The neighbors say it was because he ate the mushroom and then had intercourse with his wife.

When a family decides to make use of the mushroom they tell their friends to bring them any they see, but they ask only those whom they can trust to refrain from intercourse at that time, for if the person who gathers the mushroom has had intercourse, it will make the person who eats it crazy.

Usually it is not the sick person nor his family who eat the mushroom. They pay a 'wiseman' to eat it and to tell them what the mushroom says. (He does so with a loud rhythmic chant.) The wiseman always eats the mushroom at night, because it 'prefers to work unseen'. Usually he eats it about nine o'clock and it starts talking about a half an hour or an hour later. The Mazatecs speak of the mushroom as though it had a personality. They never say "The wiseman said the mushroom said . . ." They always quote the mushroom direct.

The wiseman always eat the mushroom raw; "If anyone cooks or burns the mushroom it will give them bad sores. "There is no specified number of how many he should eat, some wisemen eat more than others, usually they eat four or five. If he eats a lot, it 'wants to kill him'. At such a time the wiseman falls down unconscious, and comes to little by little as the other people in the room 'pray for him'. This may also happen 'if he has had intercourse too recently'.

When all goes well, the wiseman sees visions and the mushroom talks about two or three hours. "It is Jesus Christ himself who talks to us!" The mushroom tells them what made the person sick. He may say the person was bewitched; if so, he tells who did it, why, and how. He may say the person has 'fear sickness'. He may say it is a sickness curable by medicine and suggest that the person call a doctor.

More important, he will tell whether or not the person is going to live or die. If he says he will live, then "he gets better even though he has been very sick". If he says he will die, then the family start making arrangements for the funeral and he tells who should inherit his property (One of my informants admitted, however, that the mushroom occasionally makes mistakes.)

One of the "proofs" that it is "Jesus Christ himself" who talks to them is that anyone who eats the mushroom sees visions. Everyone we have asked suggests that they are seeing into heaven itself. They don't insist on that point, and as an alternative they suggest that they are seeing moving pictures of the U.S.A. Most of them agree that the wisemen frequently see the ocean and for these mountain people that is exciting.

I have asked what the wiseman looks like while under influence of the mushroom. They say that he is not sleeping, he is sitting up, with his eyes open, "awake". They say he does not drink liquor at the time, but that he may in the morning. Some of them go right out to work the next day, but some stay home and sleep "because they have been awake all night".

Although we have never been present when the mushroom was eaten, we have observed the influence it has on the people. One of our neighbors had tuberculosis and was coming to us for medical help. Then one night they called in the wiseman to eat the mushroom in his behalf. It told them that he would die.

The next day the patient no longer had any interest in our medicines; instead he began to set affairs in order for death. He quit eating solid food, restricting himself to corn gruel. About two weeks later he refused even gruel, accepting only an occasional sip of water. A few days later even water was rejected. In less than a month after he had consulted the mushroom he was dead.

Another neighboring family had a series of sickness. They consulted the mushroom for their twenty-two year old son. The mushroom said he would get better, and he did. When their eighteen-year old daughter became ill, they consulted the mushroom. It said she would get better and she did.

Then the ten year old daughter became ill. The mushroom said that this one would die. The family were amazed because her illness had not seemed serious. Of course they were grief stricken, but the mushroom said, "Don't be concerned. I'll take her soul to be with me." So, following her mother's instructions the little girl prayed to the thing talking to her, "If you don't want to cure me, take my soul." A day or two later she was dead.

Not all the Mazatecs believe that the mushroom's messages are from Jesus Christ. Those who speak Spanish and have had contact with the outside world are apt to declare, "It's just a lot of lies!" Most monolinguals, however, will either declare

hat it is Jesus Christ who speaks to them, or they will ask
ı little doubting, "What do you say, is it true that it is the
ılood of Jesus?"

I regret the survival of the use of the mushroom, for we
:now of no case in which it has had beneficial results. I wish
hey'd consult the Bible when they seek out Christ's wishes,
ınd not be deceived by a 'wiseman' and the mushrooms.

In answer to your questions:

The mushroom (called *sitho,* or affectionately *'nti sitho*) is
ırown in color and grows biggest in June and July when the
·ainfall is heaviest. At that time they may be four inches across
ınd about four inches tall. They are still plentifull in September
ınd October. By March and April, the dry season, the
nushroom is scarce, but small ones may still be found.

The mushroom grows in the grass, but when people are
ıunting for it, they look first in the places where cattle have
ɔeen, because the mushroom is most frequently found growing
ɔut of cow manure.

They do not dry the mushroom. If they cannot find one
ŗrowing, they go without. The person I asked doubted that
t was possible to dry them. At first she thought they would
ıot. Then she said that maybe they could be dried, but she
loubted that they would serve as medicine that way.

I do not know that the Mazatecs ever use the mushroom
n connection with a fiesta. For the most part it is used in connec-
:ion with sickness. I have heard of one other minor use, however.
Гhey say a man may slip a piece unto an enemy's liquor while
ıe is drinking in a saloon. If he drinks it while ceremonially
ınclean, he may go crazy. Or he might go crazy because the
nan who gathered it was ceremonially unclean. . . .

Sincerely
(Signed) Eunice V. Pike

From Valentina Pavlovna and R. Gordon Wasson. *Mushrooms. Russia and History.* 2 volumes. Pantheon Books. New York. (1957) II, pp. 242-45 (Reprinted by permission)

Cactaceae or Cactus Family

Peyote *(Lophophora Williamsii)*

ROBERT S. DE ROPP

Antiquity shrouds the origins of the cactus cult. We do not know, nor are we likely to discover, by what accident some wanderer in the Mexican deserts first stumbled upon the secret of the plant's effects. We may assume that the discovery of the drug resulted from the usual causes, a quest for food on the part of some wanderer, reduced to extremity by hunger and thirst, devouring anything containing moisture and nourishment, however evil-tasting that something might be. We can envisage that long-forgotten man, Aztec or pre-Aztec, chewing the nauseous, bitter cactus tops and lying down to rest, then, in a rising tide of astonishment, finding himself ringed on all sides with fantastic visions, with shapes, colors, odors of which he had never even dreamed. Small wonder that, when he found his way back to his tribe, he informed them that a deity dwelt in the cactus and that those who devoured its flesh would behold the world of the gods.

From Robert S. De Ropp *Drugs and the Mind.* St. Martin's Press. New York. (1957 pp. 27-28. (Reprinted by permission)

ETHEL BAILEY HIGGINS

Quiet, retiring in habit of growth as it is, and limited to a small area in its native habitat, it [the peyote] has been for centuries the subject of execration and of worship; of experiment, investigation, of study and research, and of legislation. . . What is its strong hold upon the Indian? Is it as a narcoti

Figure 1.
Flowering head of the peyote cactus *(Lophophora Williamsii)*. Photograph courtesy of R. E. Schultes.

or as a part of their religious belief? It would seem that th two blend into a composition that defies analysis; . . .

From Ethel Bailey Higgins "Peyote (*Lophophora Williamsii*) the Sacred Musroon in Religious Rites of the Indians." *Desert Plant Life.* Pasadena, Calif. 4 (1932-33 pp. 90-91.

DAVID F. ABERLE

The origin of the ritual— . . . is unknown. . . . The ritua now crystallized with minor variations as *the* peyote ritual may have arisen before the modern cult; it certainly arose either before or with it, and not subsequently. The date of origin of the cult also remains unknown. . . .

Wherever it originated, the peyote *cult* departed from earlier uses of *peyote* in its pan-Indian and nativistic character. There is some question whether it had Christian elements at this time At least by 1892 Christian elements were present . . .

Its spread in this form was rapid. A minimum of 16 tribes are documented as having the cult by 1899, and an additiona 61 had been added by 1955. . . . Peyotism reached from Alberta Manitoba, and Saskatchewan to New Mexico, Arizona, and California, and from Wisconsin to Idaho and Nevada by 1955 . . . The percentage of members varies greatly from tribe to tribe.

We will not here attempt to account for this variability, nor for the limits of spread at present. It is evident, however, that peyotism had enormous viability and appeal. It started on a small reservation in Oklahoma some 70-odd years ago and has touched at least 77 tribes—more than a tribe a year—in the United States and Canada in the interval. This appeal we have tried to locate in the *peyote* experience, in the polyvalent appea of peyotism, and in its utility for Indians in a special socio-economic niche.

It would, however, be a serious mistake to think of peyotism as merely 'spreading', like a flow of oil. It met bitter opposition from the beginning, and opposition continues today. Peyotism has been and is opposed by traditionalist Indians, modernist Indians, and whites. Broadly speaking, traditionalists oppose it as a combination of 'heathenism' and misunderstood Christianity, and as backward, and so do whites. All three oppose it on the assumption that the consumption of *peyote* has evil effects, to wit: disease, death, sexual immorality, laziness, intoxication, mental disease, malformed or stillborn infants, and addiction. To date none of these claims has been substantiated by observations under properly controlled conditions.

Opposition has taken three primary forms: talk, unofficial harassment, and legal efforts. The first efforts at suppression date from 1888. . . . There was a rush of suppressing laws in the period 1917-1923, when no less than 9 states enacted them. In recent years, however, the laws against *peyote* have been repealed in Oklahoma, Utah, Iowa, Texas and New Mexico.

So far as I know, enforcement was no more than sporadic in any of the states which have, or have had, laws: Arizona, Massachusetts, Montana, Nevada, New Mexico, North Dakota, Oklahoma, Texas, Utah, and Wyoming. . . .

From David F. Aberle *The Peyote Religion among the Navaho* . . . (c) 1966. Aldine Publishing Company. Chicago, Ill. pp. 17-18. (Reprinted by permission)

WALTER BROMBERG

The spread of its use among the American Indians has alarmed many who see in *peyote* only a narcotic drug. . . . Those in contact with the *peyote* ritual among the Indians echo Dr. Weir Mitchell's sentiment expressed in 1896, after laboratory experimentation with thc alkaloid of the *peyote* plant:

> "I predict a perilous reign of this mescal habit when this agent becomes obtainable. The temptation to call again the enchanting magic will be too much for some men to resist after they have set foot in this land of fairy colors."

From Walter Bromberg, M.D. "Storm over Peyote". *Nature Magazine* (American Nature Association) Washington. etc. 35 (1942) p. 410.

WESTON LA BARRE

One of the most important and striking of . . . [Peyote] uses is in prophecy and divination. We find the Spanish missionaries in Mexico early protesting against this abomination. Padre Arlegui [in his *Crónica* . . . México, 1737] after mentioning the therapeutic uses to which the Zacatecans put *peyote,* complains that

> "this would not be so bad if they did not abuse its virtues, for, in order to have a knowledge of the future and find out how their battles will turn out, they drink it brewed in water, and, as it is very strong, it intoxicates them with a paroxysm of madness, and all the fantastic hallucinations that come over them with this horrible drink they seize upon as omens of their future, imagining that the root has revealed to them their future."

From Weston La Barre "The Peyote Cult" Yale University *Publications in Anthropology* #19 Yale University Press, New Haven, Conn. (1938) p. 23. (Reprinted by permission)

BERNARDINO DE SAHAGUN

"On Certain Intoxicating Herbs."

. . . There is another herb, like native tunas [prickly pears]; it is called *péyotl;* it is white; it grows in the north region [called Mictlán]. Those who eat or drink it see visions either frightful or mirthful; the intoxication lasts two or three days and then ceases. It is a common food of the Chichimecas, for it sustains them and gives them courage to fight and not to feel hunger nor thirst; and they say that it protects them from all dangers.

From Fray Bernardino de Sahagún (1499-1590) *Historia General de las Cosas de Nueva España*. (Translation: General History of Things of New Spain (Mexico).) 5 volumes. Pedro Robredo. México, D.F. (1938) Book XI, ch. vii.

FRANCISCO HERNANDEZ

"Of the *Péyotl* of Zacateco, a Woolly and Delicate Root."

It is a medium sized root which does not grow branches

or leaves above the ground, but only adheres to the ground with some wooliness. For this reason I was not able to depict it. They assert that it does harm to both males and females. It has a sweet taste and is moderately hot. The Indians contend that, when crushed and applied, it cures pains in the joints. Miraculous properties are attributed to this root (if any faith may be given to what they commonly say among themselves). To wit, that those who eat it will be able to foresee and predict everything, such as whether they should attack the enemy on the following day or rather wait for favorable times; or who had stolen a utensil, or other matters of like nature which the Chichimecas attempt to find out with the help of this plant. Furthermore, if they wish to know where they would find this root hidden in the ground, they eat another one of them and would find the place. It grows in humid places rich of calcium.

From Dr. Francisco Hernandez (1515-1587), personal physician to the Emperor. *Opera* . . . (Translation: Works) 3 vols. Printed by Ibarra's Heirs. Madrid. (1790) III., pp. 70/71. (Book XV, ch. xxv.)

JOSÉ DE ORTEGA

[On the Superstitious Rites of the Nayarits.]

In September, when the maize starts to form grains, the Indians, unless driven by want, did not taste the maize before it was blessed by the priests in the temples of their gods. They had a special rite for this purpose: The fruits that each Indian brought were placed on a tree trunk next to all Indians and their families. Two Indians stood watch at either side of the trunk lest some youngsters might be tempted to taste the fruits; for if someone ate these fruits before they were blessed—and everyone feared that the youngsters might do this either by oversight or from mischievousness—the gods would punish them all by afflicting them with the discomfort of herpes. Near the trunk was seated the player of the fiddle, consisting of a deep trough with a string fastened to it, which he struck with a small stick. This produced such a harmony that probably it would not have offended the ear of the listener if the confused humming of the singers had not disturbed the music. Close to the musician was seated the choirmaster who had the task of keeping time. Each had an assistant to take his place should he become tired. Nearby was placed a tray filled with *peyote* or the devil's root. They drunk the ground-up root, so that they might not become weakened by the severe damage of a performance of so long duration. This began by forming a circle of as many men and women as would fill the space which had been swept off for this purpose. One after the other went dancing into this circle or kept time with his feet, keeping in the middle the musician and the choirmaster whom they had invited, and singing in the same cacographic tune that he set them. They danced from five o'clock in the afternoon to seven o'clock in the morning, without stopping or leaving the circle. When the dance was ended, all who could hold themselves on their feet stood; for the majority from the *peyote* and the wine, were unable to use their legs to keep themselves upright or even to take notice of the blessing the high priest gave to the fruits. . . .

From [Padre José de Ortega, S.J.](1700-1768) *Apostolicos afanes de la Compañia de Jesús en su Provincia de México.* (Translation: Apostolic solicitudes of the Society of Jesus in their Province of Mexico.) P. Nadal. Barcelona, (1754) Book I, ch. iii.

HEINRICH KLÜVER

As a second example of mescal visions, we quote from the report [1896] of [Silas] Weir Mitchell [1829-1914] who 'at 12 noon of a busy morning' took 11/2 drachms of an extract 'of which each drachm represented one mescal button.' One hour hereafter, little over a drachm was taken and at about four o'clock half an ounce of this extract in three doses.' Soon Mitchell found himself 'deliciously at languid ease.' At 5.40, he noticed a number of star points and fragments of stained glass with closed eyes. He went into a dark room: 'The display which for an enchanted two hours followed was such as I find it hopeless to describe in language which shall convey to others the beauty and splendour of what I saw.' 'Stars . . . delicate floating films of colour . . . then an abrupt rush of countless points of white light swept across the field of view, as if the unseen millions of the Milky Way were to flow a sparkling river before the eye . . . zigzag lined of very bright colours . . . the wonderful loveliness of swelling clouds of more vivid colours gone before I could name them . . .' Then, for the first time, 'definite objects associated with colours' appeared. 'A white spear of grey stone grew up to huge height and became a tall, richly finished Gothic tower of very elaborate and definite design, with many rather worn statues standing in the doorways or on stone brackets. As I gazed, every projecting angle, cornice and even the faces of the stones at their joinings were by degrees covered or hung with clusters of what seemed to be huge precious stones, but uncut, some being more like masses of transparent fruit. These were green, purple, red and orange; never clear yellow and never blue. All seemed to possess an interior light and to give the faintest idea of the perfectly satisfying intensity and purity of these gorgeous colour-fruits is quite beyond my power. All the colours I have ever beheld are dull as compared to these. As I looked, and it lasted long, the tower became of a fine mouse hue, and everywhere the vast pendant masses of emerald green, ruby reds and orange began to drip a slow rain of colours. . . . After an endless display of less beautiful marvels, I saw that which deeply impressed me. An edge of a huge cliff seemed to project over a gulf of unseen depth. My viewless enchanter set on the brink a huge bird claw of stone. Above, from the stem or leg, hung a fragment of some stuff. This began to unroll and float out to a distance which seemed to me to represent Time as well as immensity of Space. Here were miles of rippled purples, half transparent, and of ineffable beauty. Now and then soft golden clouds floated from these folds, or a great shimmer went over the whole of the rolling purples, and things, like green birds, fell from it, fluttering down into the gulf below. Next, I saw clusters of stones hanging in masses from the claw toes, as it seemed to me miles of them, down far below into the underworld of the black gulf. This was the most distinct of my visions.' In his last vision, Mitchell saw the beach of Newport with its rolling waves as 'liquid splendours huge and threatening, of wonderfully pure green, or red or deep purple, once only deep orange, and with no trace of foam. These water hills of colour broke on the beach with myriads of lights of the same tint as the wave.' Again the author considers it totally impossible to find words to describe these colours. 'They still linger visibly in my memory and left the feeling that I had seen among them colours unknown to my experience.'

From Heinrich Klüver *Mescal and Mechanisms of Hallucinations.* (c) 1966. University of Chicago Press. Chicago & London, pp. 15-17. (Reprinted by permission of the Orthological Institute, London)

HAVELOCK ELLIS

On Good Friday, I found myself entirely alone in the quiet rooms in the Temple which I occupy when in London and judged the occasion a fitting one for a personal experiment. I made a decoction . . . of three [mescal] buttons, the full physiological dose, and drank this at intervals between 2.30 and 4.30 p.m. The first symptom observed during the afternoon was a certain consciousness of energy and intellectual power *(*)*. This passed off, and about an hour after the final dose I felt faint and unsteady; the pulse was low, and I found it pleasanter to lie down. I was still able to read, and I noticed that a pale violet shadow floated over the page around the point at which my eyes were fixed. I had already noticed that objects not in the direct line of vision, such as my hands holding the book, showed a tendency to look obtrusive, heightened in color, almost monstrous, while, on closing my eyes, afterimages were vivid and prolonged. The appearance of visions with closed eyes was very gradual. At first there was merely a vague play of light and shade which suggested pictures, but never made them. Then the pictures became more definite, but too confused and crowded to be described, beyond saying that they were of the

* I pass lightly over the purely physiological symptoms which I have described in some detail in a paper on "The Phenomena of Mescal Intoxication." (*Lancet,* June 5, 1897), which, however, contains no description of the visions.

same character as the images of the kaleidoscope, symmetrical groupings of spiked objects. Then, in the course of the evening, they became distinct, but still indescribable—mostly a vast field of golden jewels, studded with red and green stones, ever changing. This moment was, perhaps, the most delightful of the experience, for at the same time the air around me seemed to be flushed with vague perfume—producing with the visions a delicious effect—and all discomfort had vanished, except a slight faintness and tremor of the hands, which, later on, made it almost impossible to guide a pen as I made notes of the experiment; it was, however, with an effort, always possible to write with a pencil. The visions never resembled familiar objects; they were extremely definite, but yet always novel; they were constantly approaching, and yet constantly eluding, the semblance of known things. I would see thick, glorious fields of jewels, solitary or clustered, sometimes brilliant and sparkling, sometimes with a dull rich glow. Then they would spring up into flowerlike shapes beneath my gaze, and then seem to turn into gorgeous butterfly forms or endless folds of glistening, iridescent, fibrous wings of wonderful insects; while sometimes I seemed to be gazing into a vast hollow revolving vessel, on whose polished concave mother-of-pearl surface the hues were swiftly changing. I was surprised, not only by the enormous profusion of the imagery presented to my gaze, but still more by its variety. Perpetually some totally new kind of effect would appear in the field of vision; sometimes there was swift movement, sometimes dull, somber richness of color, sometimes glitter and sparkle, once a startling rain of gold, which seemed to approach me. Most usually there was a combination of rich, sober color, with jewel-like points of brilliant hue. Every color and tone conceivable to me appeared at some time or another. Sometimes all the different varieties of one color, as of red, with scarlets, crimsons, pinks, would spring up together, or in quick succession. But in spite of this immense profusion, there was always a certain parsimony and aesthetic value in the colors presented. They were usually associated with form, and never appeared in large masses, or if so, the tone was very delicate. I was further impressed, not only by the brilliance, delicacy, and variety of the colors, but even more by their lovely and various textures—fibrous, woven, polished, glowing, dull veined, semitransparent—the glowing effects, as of jewels, and the fibrous, as of insects' wings, being perhaps the most prevalent. Although the effects were novel, it frequently happened, as I have already mentioned, that they vaguely recalled known objects. Thus, once the objects presented to me seemed to be made of exquisite porcelain, again they were like elaborate sweetmeats, again of a somewhat Maori style or achitecture; and the background of the pictures frequently recalled, both in form and tone, the delicate architectural effects as of lace carved in wood, which we associate with the *moucharabieh* work of Cairo. But always the visions grew and changed without any reference to the characteristics of those real objects of which they vaguely reminded me, and when I tried to influence their course it was with very little success. On the whole, I should say that the images were most usually what might be called living arabesques. There was often a certain incomplete tendency to symmetry, as though the underlying mechanism was associated with a large number of polished facets. The same image was in this way frequently repeated over a large part of the field; but this refers more to form than to color, in respect to which there would still be all sorts of delightful varieties, so that if, with a certain uniformity, jewel-like flowers were springing up and expanding all over the field of vision, they would still show every variety of delicate tone and tint.

Weir Mitchell found that he could only see the visions with closed eyes and in a perfectly dark room. I could see them in the dark with almost equal facility, though they were not of equal brilliancy, when my eyes were wide open. I saw them best, however, when my eyes were closed, in a room lighted only by flickering firelight. This evidently accords with the experience of the Indians, who keep a fire burning brightly throughout their mescal rites.

The visions continued with undiminished brilliance for many hours, and as I felt somewhat faint and muscularly weak, I went to bed, as I undressed being greatly impressed by the red, scaly, bronzed, and pigmented appearance of my limbs whenever I was not directly gazing at them. I had not the faintest desire for sleep; there was a general hyperaesthesia of all senses as well as muscular irritability, and every slightest sound seemed magnified to startling dimensions. I may also have been kept awake by a vague alarm at the novelty of my condition, and the possibility of further developments.

After watching the visions in the dark for some hours I became a little tired of them and turned on the gas. Then I found that I was able to study a new series of visual phenomena, to which previous observers had made no reference. The gas jet (an ordinary flickering burner) seemed to burn with great brilliance, sending out waves of light, which expanded and contracted in an enormously exaggerated manner. I was even more impressed by the shadows, which were in all directions heightened by flushes of red, green, and especially violet. The whole room, with its white-washed but not very white ceiling, thus became vivid and beautiful. The difference between the room as I saw it then and the appearance it usually presents to me was the difference one may often observe between the picture of a room and the actual room. The shadows I saw were the shadows which the artist puts in, but which are not visible in the actual scene under normal conditions of casual inspection. I was reminded of the paintings of Claude Monet, and as I gazed at the scene it occured to me that mescal perhaps produces exactly the same conditions of visual hyperaesthesia, or rather exhaustion, as may be produced on the artist by the influence of prolonged visual attention. I wished to ascertain how the subdued and steady electric light would influence vision, and passed into the next room; but here the shadows were little marked, although walls and floor seemed tremulous and insubstantial, and the texture of everything was heightened and enriched.

About 3.30 a.m. I felt that the phenomena were distinctly diminishing—though the visions, now chiefly of human figures, fantastic and Chinese in character, still continued—and I was able to settle myself to sleep, which proved peaceful and dreamless. I awoke at the usual hour and experienced no sense of fatigue nor other unpleasant reminiscence of the experience I had undergone. Only my eyes seemed unusually sensitive to color, especially to blue and violet; I can, indeed, say that ever since this experience I have been more aesthetically sensitive than I was before to the more delicate phenomena of light and shade and color.

From Havelock Ellis (1859-1939) "Mescal: A New Artificial Paradise." *The Contemporary Review.* A. Strahan, etc., London. 73 (1898) pp. 131-34.

PAUL RADIN

"John Rave's Account of the Peyote Cult and His Conversion."

During 1893-94, I was in Oklahoma with *peyote* eaters.

In the middle of the night we were to eat *peyote.* We ate it and I also did. It was the middle of the night when I got frightened, for a live thing seemed to have entered me. "Why did I do it?" I thought to myself. I should not have done it, for right at the beginning I have harmed myself. Indeed, I should not have done it. I am sure it will injure me. The best thing will be for me to vomit it up. Well now, I will try it. After a few attempts I gave up. I thought to myself, "Well, now you have done it. You have been going around trying everything and now you have done something that has harmed you. What is it? It seems to be alive and moving around in my stomach. If only some of my own people were here! That would havè been better. Now no one will know what has happened to me. I have killed myself."

Just then the object was about to come out. It seemed almost out and I put out my hand to feel it, but then it went back again. "O, my, I should never have done it from the beginning. Never again will I do it. I am surely going to die."

As we continued it became day and we laughed. Before that I had been unable to laugh.

The following night we were to eat *peyote* again. I thought to myself, "Last night it almost harmed me." "Well, let us do it again," they said. "All right, I'll do it." So there we ate seven *peyote* a piece.

Suddenly I saw a big snake. I was very much frightened.

Then another one came crawling over me. "My God! where are these coming from?" There at my back there seemed to be something. So I looked around and I saw a snake about to swallow me entirely. It had legs and arms and a long tail. The end of this tail was like a spear. "O, my God! I am surely going to die now," I thought. Then I looked again in another direction and I saw a man with horns and long claws and with a spear in his hand. He jumped for me and I threw myself on the ground. He missed me. Then I looked back and this time he started back, but it seemed to me that he was directing his spear at me. Again I threw myself on the ground and he missed me. There seemed to be no possible escape for me. Then suddenly it occurred to me, "Perhaps it is this *peyote* that is doing this thing to me?" "Help me, O medicine, help me! It is you who are doing this and you are holy! It is not these frightful visions that are causing this. I should have known that you were doing it. Help me!" Then my suffering stopped. "As long as the earth shall last, that long will I make use of you, O medicine!"

This had lasted a night and a day. For a whole night I had not slept at all.

Then we breakfasted. Then I said, when we were through, "Let us eat *peyote* again to-night." That evening I ate eight *peyote*.

In the middle of the night I saw God. To God living up above, our Father, I prayed. "Have mercy upon me! Give me knowledge that I may not say and do evil things. To you, O God, I am trying to pray. Do thou, O Son of God, help me, too. This religion, let me know. Help me, O medicine, grandfather, help me! Let me know this religion!" Thus I spoke and sat very quiet. And then I beheld the morning star and it was good to look upon. The light was good to look upon. I had been frightened during the night but now I was happy. Now as the light appeared, it seemed to me that nothing would be invisible to me. I seemed to see everything clearly. Then I thought of my home and as I looked around, there I saw the house in which I lived far away among the Winnebago, quite close to me. There at the window I saw my children playing. Then I saw a man going to my house carrying a jug of whisky. Then he gave them something to drink and the one that had brought the whisky got drunk and bothered my people. Finally he ran away. "So, that is what they are doing." I thought to myself. Then I beheld my wife come and stand outside of the door, wearing a red blanket. She was thinking of going to the flagpole and was wondering which road she should take. "If I take this road I am likely to meet some people, but if I take the other road, I am not likely to meet anyone."

Indeed, it is good. They are all well—my brother, my sister, my father, my mother. I felt very good indeed. O medicine, grandfather, most assuredly you are holy! All that is connected with you, that I would like to know and that I would like to understand. Help me! I give myself up to you entirely!

For three days and three nights I had been eating medicine and for three days and three nights I had not slept. Throughout all the years that I had lived on earth, I now realized that I had never known anything holy. Now, for the first time, I knew it. Would that some of the Winnebagoes might also know it! . . .

From Paul Radin (1883-1959) "The Winnebago Tribe." U. S. Bureau of American Ethnology. *Thirty-Seventh Annual Report of the Board of Regents of the Smithsonian Institute to the Secretary. (1915-16)* Government Printing Office. Washington. (1923) pp. 389-91.

F. H. DAIKER

. . . . two affidavits submitted by Indians who are partakers of *peyote:*

"He is the head of this religion on the reservation and he has twelve men who are his apostles. The best six of them sit on his right hand whenever they have meetings, and the six who are not so much good on his left hand. The apostles are dressed all in white and sit all night as a penance; They have a fire in the centre of their circle and burn something like pine cones or leaves, and the smoke curls up like a sacrifice. He does not believe in having the Bible in his ceremony. They eat the mescal or *peyote,* which are the same thing, and rattle gourds and drum and sing. Before they sing they pass the *peyote* around. They begin taking this medicine along about dark, and when they pass it, ask you how many you want, and they often try to persuade you to take more than you want. The medicine does not work right away, but after it begins to take effect along toward midnight they begin to cry and sing and pray and stand and shake all over, and some of them just sit and stare. I used to sit in their range right along, and ate some of their medicine, but after I ate it the first time I was kind of afraid of it. It made me feel kind of dizzy and my heart was kind of thumping and I felt like crying. Some of them told me that this was because of my sins. It makes me nervous, and when I shut my eyes I kind of see something like an image or visions, and when my eyes are open I can't see it so plain. One of these fellows took 12 beans, or 12 *peyote,* sitting with some girls. They asked me if I would read the Bible. One of the sides of the religion believe in having the Bible read. I told them that I could not see it because my eyes were kind of dizzy. The Indians began to stand up around me and to cry. After I had taken 12 beans or *peyote* I saw a mountain with roads leading to the top and people dressed in white going up these roads. I got very dizzy and I began to see all kinds of colors and arrows began to fly around me. I began to perspire very freely. I asked to be taken out of doors. At that time it was twenty below zero. I felt better when I got out of doors. When I went in again I began to hear voices just like they came from all over the ceiling and I looked around in the other room and thought I heard women singing in there, but the women were not allowed to sing in the meetings usually, and so this was kind of strange. This was about the first time I had eaten the *peyote* and I was scared, and I asked some of the people what kind of stuff this was I had been eating. I was afraid at first to eat the *peyote* but afterwards I kind of did it because they were doing it, and they were my friends, and I thought I liked to try it. The leader of this part of the *peyote* leaders told me that this *peyote* was the Holy Ghost, but I told him that that could not be so and it was wrong of him to say that. I kind of made up my mind that this thing (the eating of *peyote*) was wrong and there was more to it than I thought; so I decided to try it out fully and see really what there was in it. My mother is a strong Christian woman, and although I was not a church member, I believed in prayer; so I prayed to God to help me, and I made up my mind that I was willing to sacrifice my life to expose this bad religion and I started out by eating 36 of these *peyote.* They assumedly took effect. I got just like drunk, only more so and I felt kind of good, but more good than when I drink whiskey, and then after that I began to see a big bunch of snakes crawling all around in front of me and it was a feeling like as if I was cold came over me. The treasurer of the sacred *Peyote* Society, which is the name by which they call this Mescal religion or *Peyote* Church, was sitting near me, and I asked him if he heard young kittens. It sounded as if they were right close to me, and then I sat still for a long time and I saw a big black cat coming toward me, and I felt him just like a big tiger walking up on my legs towards me, and when I felt his claws I jumped back and kind of made a sound as if I was afraid, and he asked me to tell him what was the matter, so I told him after a while. I did not care to tell at first, but I made up my mind then, after what I saw, that I would not take another one of these *peyotes* if they gave me a ten dollar bill. I was sick to my stomach and trembled all over, and the next day and for several days after I could not dress and had chills and a headache and was awful sick. In this sacred *Peyote* Society they have a form of baptism and they baptise with the tea made from stewing the *peyote*, and they baptise in the name of the Father and the Son and the Holy Ghost, the Holy Ghost being the *peyote*. Then you drink some of the tea and they make signs on your forehead with the tea and then take an eagle's wing and fan you with it. I heard an educated Indian and he said in a meeting on Sunday morning, My friends, I am glad I can be here and worship this medicine *(peyote)* with you and we must organize a new church and have it run like the "Mormon Church."

"I did not know what the Indians were doing some of the

time and could not understand their language, but one time they baptized me with mescal tea or *peyote* tea, and gave me the name of John White Eagle. They say that if you eat this bean it will cure you from drinking whiskey or other intoxicants, and it makes you saving and a better worker. I know that this is not true for I have seen one member who belongs to the *Peyote* Society drunk in Sioux City on whiskey and beer, but he is still using the mescal or *peyote*. Most all the mescal or *peyote* eaters go off and get drunk occasionally. There are a few who can't get drunk enough to suit them on the *peyote*. I have been to Sioux City with many of the Mescal or *Peyote* Society and got drunk with them. This is a common thing for them to do, because I have seen them do it. So when they say that it keeps them from drinking I know that that is a lie. I think maybe when they first start to use *peyote* they give up whiskey for a little while, but they soon want it again. They are also spending a great deal of their money on these meetings and to make the Society grow. Many of the mescal or *peyote* eaters use from sixty to one hundred dollars a night and mortgage their things to get money to feed the meeting. I think the leaders spend much of the money themselves. I know they collected lots of money to build a church and I gave most of my wages to them, but when they tried to find out what the treasurer did with the money it was all gone, so I would not give any more money. One of the leaders has sold a great amount of his land and it had all been spent to feed the meetings, or nearly all of it. One Indian who owed me some money gave the money to the leader of this Sacred *Peyote* Society, but I never saw a cent of it. I could see very plainly that the Indian leaders of this *peyote* religion are making their living off the ignorant Indians, so I know that it does not make them thrifty, and in fact if they keep on the way they are doing, it will not be very long until they are very poor. I am sure if you would look at the Agency records or ask the Agent he could tell you that that was so. Another thing that is very bad is that when a baby is born they give it some of this mescal or *peyote* tea to drink and this kills them. There are many babies die from that cause. One Indian's baby was given some mescal or *peyote* tea and died; he told me he gave it lots of medicine, by which he meant the mescal or *peyote* tea, and I told him that was very wrong. Not only the babies but the sick people are given this mescal tea. They bring them into a house or their meeting place and all gather round him and feed him mescal or *peyote*. Last winter an Indian was carried into the mescal or *peyote* church and they kept him full of this mescal or *peyote* tea for about four days and nights, and then he staggered around the church rising to his feet and prayed and bid goodby to his friends and said, 'I am going to Jesus' and dropped down dead. The wife of one Indian was given too many doses of this *peyote* tea and died. A big strong man used lots of mescal or *peyote* and whiskey and died in two days. Another one used a lot of this mescal or *peyote* and he lost all his teeth. I am very sure that if I had kept on taking this mescal or *peyote* I would have died because I nearly died as it was. One time when I ate 36 beans I just felt as I could throw my arms out and my arms left me, went off in the air, and I felt I was all going to pieces. Everybody that I saw looked so much larger; their faces were large. I remember one time it was said in my hearing, 'I saw Jesus' picture in the bean soup' meaning the mescal or *peyote* tea. Whenever they pray in the meeting, they put the bean on a white cross or white napkin on the ground and they touch the bean first and then touch their lips and then pray to the mescal or *peyote* and then to God. They consecrate themselves to the medicine or the mescal or *peyote*. Whenever they eat these beans it makes you feel more whatever you are thinking about, (intensifies your thought). One man ate 75 beans or *peyote* and it killed him so they reduced the dose to 20 or 30 *peyote* or mescals. They are now getting afraid of that I think. Another thing that I noticed is that many are getting blind who use this medicine very much. When I was eating it, I just saw flames shooting out from my eyes, and I could not sleep or close my eyes. It kept me awake, affected my nerves, made me nervous."

"The meetings are very immoral. I do not think I could

tell all I have seen in writing, but I have seen many very bad things, That is, they look bad."

From F. H. Daiker "Liquor and *Peyote* a Menace to the Indians." *Thirty Second Report of the Annual Lake Mohonk Conference on the Indians and other Dependent Peoples.* Philadelphia. etc. (October 1914) pp. 66, 67 & 68.

HENRI MICHAUX

[The French painter-poet describes his mescaline experience:]

Drugs weary us with their paradises. Let them enlighten us a little instead. Ours is not a century of paradises.

"Mescaline and Music."

. . . In mescaline intoxication the rhythms, indeed, are very frequently felt. It is even surprising that they remain so independent of the music, that they never attach themselves to it, or hardly ever, or only imperfectly. I myself had experienced something of an extremely hybrid nature and I was waiting to speak of it for a more conclusive adventure, which has never occurred. I shall therefore say a few words about it here. It was in 1956, in the course of a kind of 'erotic' trance. What had chiefly happened to me on that day was orgiastic and fantastic visions and rhythms of the same order. Suddenly songs came out—yes, came out. For as much as I heard them, I felt them coming out, having to come out, in a hurry to come out, incoercible phonating movements, which surely came from choral singers whose voices I heard, but also had their origin in my throat, which was possessed by a kind of vocal urge which made me a coparticipant and active.

And what was it? Passages from Olivier Messiaen's *Trois Petites Liturgies de la présence divine.* But in what state! Passages, bits rather, and bits which might have been picked out by a man at the peak of exasperation, who cannot stand sounds more than fifteen seconds in succession, but who comes back to them again and again, ever in the same state of nerves, ever in the same madly headlong fashion. Its fragments were so hurried in tempo that I expected to hear them jumbled, but no, the buffoonish burgeoning proceeded, without a mistake, in spite of the unbelievable speed, in spite of outbursts of notes like hurried evacuations, like machine-gun fire. Overflowing and the next moment stopping, the music went on and on, deformed, the negation of music, the negation of mystical theology. Never would I have thought any music capable of becoming so shameless, bawdy, dissolute, mad, impious, ignoble, subversive. There were also at moments snatches, pranks, high-perched notes, inept couplings, never-heard musical frictions, extravagant *abbellimenti,* a mad partitioning, voices strangely coupled with barks or jungle howls. Few inventions on the whole in the structures, but for parody anything will serve. Everything strikes goal.

The pieces, cut out from the work, were beginning to be themselves crossed by other bits of music, some of them highly syncopated, others not, picked up from old reminiscences suddenly reawakened, strange bits, bit upon bit. 'Uniquely through love' in particular has set off the system of bits fitting in so well with mescaline, with its mechanical side. Bravura pieces, of the type of the great arias from *La Tosca,* burst forth, dazzling and inept, shards of abhorred tunes I must once have heard with nausea and shame played in the street by some barrel organ, a derision and degradation of droll and rococo musical sentiment, but the chants remained the *pièce de résistance* (what resistance!) which struggled against the demoniac hodgepodge, which was trying to carry it away and periodically ruined its effects. A Gargantuan laughter, which I was unable to achieve, might have liberated me. But profanation blanketed the whole. Moreover, the music would be there and suddenly would stop, and this made no great difference. This was odd. Often I would catch myself following it without hearing it, thinking I was still hearing it when I realized it had 'gone blank', in other words that no sound was coming. But my unchanged trance continued without it and it was when it resumed, by those sounds which suddenly burst forth and attacked me, that I again became aware of it. Tens of seconds, perhaps more, had flowed by, the music having disembarked, before it returned, re-embarking with a great din. The vocal side remained dominant, the instrumental merely following. (Which is surprising, as I like only instruments and, I might almost say, never voices. But what you detest is stronger, often more fixed in you than what you have loved, which has given you no trouble.) Ill-treated, perverted, ridiculed and ridiculing, this music-outraging music had flights that the greatest lyrics do not have. Even truncated, even vilified, even punctuated by disasters, it was in no way abject. An ignoble pleasure was its center, its nature, its secret, an omnipresent, spasmodic, unendurable pleasure. On hearing it, on following it, you were subjected to pangs, to slashings, to decomposed expansions, to collapses and to rendings. Every tutelage, every protection, every musical propriety having been thrown out of the window, you were on a non-physical bed raked by pleasurable indulgence. A subversive carnival, ejaculations of joy, composed of enjoyments which were like crashes, like defenestrations, a maddening exasperation which nothing could ever appease. In a frenzy of liberation, while melodies timbled down, others were intercepted, seized and raped, as it were, then savagely tossed aside. Insane ritornellos, hurriedly pieced together, many-voiced songs, each fanning out in a harum-scarum fashion, making excruciating rips in the fabric of sound.

The immense cataracts of a very large river, which should happen at the same time to be the enormous body in the throes of pleasure of a recumbent giantess with a thousand love cracks, inviting and dispensing love, would have been something similar.

But it was music, more insatiable than any monster, the possessed music of the mescalinian demon, given over to its devastations, to its twists and turns, and giving over to them.

From Henri Michaux (1899-) *Light Through Darkness.* Translated by Haakon Chevalier. (c) Editions Gallimard, 1961. Translation (c) The Orion Press, 1963 pp. 57-59. (Reprinted by permission of Grossman Publishers and Librairie Gallimard, France)

PETER T. FURST

The Primordial Peyote Quest

> This comes to us from ancient, ancient times. The times of my great-great-grandfathers, those who were the fathers of my great-grandfather, fathers of my grandfather who was the *mara'akáme* [or shaman-priest and singer], fathers of my father. This is a story from those very ancient times. . . . Those ancient ones of whom I speak, they began to say to one another, 'How will it turn out well, so that there will be unity of all, this unity we have?' And another said, 'Ah, that is a beautiful thing, that which is our life. It is the *hikuri* [peyote].' And another said, 'It is like a beautiful flower, as one says. It is like the Deer. It is our life. We must go so that it will enable us to see our life.'

So begins an account by my long-time Huichol friend Ramón Medina Silva of the original journey to *Wirikúta*—the primordial quest of the gods that provides the mythological model for the Huichol peyote pilgrimage.

According to the myth, the ancient gods had come together in the first *túki,* the prototypical Huichol sanctuary constructed by *Tatewari,* so that each might have his proper place. When they met together they discovered that all were ill—one suffered a pain in his chest, another in his stomach, a third in his eyes, a fourth in his legs, and so forth. Those responsible for rain were giving no rain; those who were masters of animals were finding nothing to hunt. It was a time of general malaise in the Sierra, and none knew how to "find his life."

Into this assembly of the ailing gods entered the *Mara'akáme, Tatewari,* tutelary deity of Huichol shamans. It was *Tatewari* who had called them together, as the singing shaman of the temple to this day calls the supernaturals together "to take their proper places." "What can be ailing us?" they asked, and each spoke of his infirmities. "How shall we be cured? How shall we find our life?"

Tatewari told them that they were ill because they had not gone to *Wirikúta* (Réal de Catorce), the sacred land of the peyote, the place to the east where the Sun was born. If they wished to regain their health, they must prepare themselves ritually and follow him in their proper order on the long and difficult journey to the peyote. They must fast and touch neither salt nor *chile.* No matter how hungry or thirsty they became, they must nibble only dried tortillas and assuage their thirst with but a drop or two of water.

And so he placed them in their proper order, one after the other. No females were present—they would join the men later, at the sacred lakes or water holes called *Tatéi Matiniéri* (Where Our Mother Dwells), which lie within sight of the sacred mountains of *Wirikúta.*

Not all the divine peyote seekers completed the primordial quest. Some, like Rabbit Person, and Hummingbird Person, were forced by hunger, thirst, or sheer exhaustion to leave the ritual file. They remained behind in their animal form in places which became sanctified by their presence and which, like the other stopping places of the divine pilgrims, were forever after acknowledged with votive offerings and prayer by those who journey to the peyote. But the principal male gods and the female ones—the Rain Mothers and those of the Earth-Ready-for-Planting and of fertility and children—they followed *Tatewari* to the sacred mountains at the end of the world—"to the fifth level"—where the Deer-Peyote revealed itself to them in the ceremonial hunt. In this way they "found their life" and by their example taught the Huichol how to attain this.

Peyote pilgrimages may take place at any time between the end of the rainy season, in October-November, and early spring. . . .

From Peter T. Furst (Professor of Anthropology and Chairman of the Department of Anthropology at the State University of New York at Albany.) Article entitled "To Find Our Life: Peyote among the Huichol Indians of Mexico". From *Flesh of the Gods,* edited by Peter T. Furst. (c) 1972 by Praeger Publishers, Inc., New York. pp. 145-146. (Reprinted by permission)

Achuma *(Trichocereus pachanoï* BRIT. & ROSE)

BERNABÉ COBO

The *achuma* is a kind of thistle [that is a cactus]. [The name of this plant probably derived from the word *chumarse* or getting drunk.] . . . It grows as tall as a man, sometimes taller; it is as thick as a man's leg, square, and has the same color as the yellow aloe. . . . With this plant the demon kept the Indians of Peru mislead in their idolatries; they used this plant for their lies and superstitions. Those who drink the juice of this plant, lose their senses and remain as if they were dead and it sometimes happened that some of them died because of the coldness which their brains received. Enraptured by this drink, the Indians dreamed of a thousand extravagances believing these to be true. . . .

From Padre Bernabé Cobo (S. J.) (1582-1657) *Historia Natural y Moral de las Indias.* (Translation: Natural and Moral History of the Indias.) (Written in 1653). 4 Volumes. E. Rasco. Sevilla. 1890-93 (First edition) Book V, ch. vii, (I, 451).

San Pedro *(Trichocereus pachanoï)*

MARLENE DOBKIN

In north coastal Peru, a mescaline cactus called *San Pedro (Trichocereus pachanoï)* is employed by folk healers in the diagnosis and cure of illness. . . .

Although the Peruvian coastal peasant recognizes empirical etiology of disease, ultimate cause of illness is a supernatural one. Empirical cures, although valued, are secondary or supplementary to the utilization of magical and supplicatory techniques. A major etiological category of illness is *mal aires,* which encompasses disease-producing vapors or airs believed to emanate from tombs or ruins of sacred places. [In this coastal Mestizo villages some 500 miles north of Lima] pains in various parts of the body, paralysis of extremities, pneumonia and gastro-intestinal disturbances are often attributed to *mal aires*. . . .

Although all healers employ *Trichocereus pachanoï* in their curing ceremonies, variations in additives to the potion as well as ritualistic differences exist. Some healers add *Condorillo (Lycopodium* sp.), *Misha (Datura arborea)* or *Hornamo* (unidentified) which increases both the purgative and psychodyspeptic effect. I was told by one healer that too large a dose of *Misha* could result in death to the patient, especially if he were already weakened by illness. . . . Generally, the cactus is cut up into pieces and boiled several hours in a large tin of water until only the essence remains. . . .

The *materia medica* and *magica* of the healers varies, although distinct categories of objects can be delineated in most healing sessions. Since tables are laid to bewitch as well as to cure, the *mesa* of a folk healer contains not only the famed potion, to be drunk and medicinal herbs, but charms and defenses against counter-magic as well. The largest single category present are highly polished and oddly shaped stones, each one of which has its own name. The curers believe that these stones adopt the form of persons and animals which attack enemies. Each stone is believed able to receive orders from the curer. Several of the stones are called "herb stone", because the curer, while under the influence of the drug, receives a vision to help him the right herb to prescribe.

Another major category are defenses—consisting of a series of highly polished wooden sticks placed at the head of the table, in a protective file. At least one of these is an elegantly carved sword. The defenses are believed to keep evil spirits from approaching. Placed below the sword and sticks are various religious statues and holy pictures, such as the Virgin of Mercy, the Cross of Chalfon, religious pictures and crucifixes. [Etc.] . . .

Under a thatched, open-air shelter located some distance from his farm house, the curer assembled three patients who had come to him . . . The ceremony took place at night and was conducted without any artificial light. The *materia medica* and *magica* . . . was laid out on a cloth, with the patients seated along one side of it. Don M. had an apprentice . . . to assist him in the ceremony. A mixture of tobacco and water was inhaled from a shell as a nasal snuff by the assistant. At the same time, air was gulped to help bring the liquid snuff more quickly into the lungs. The *maestro* began to sing rather melodic, lulling songs, which began with the Lord's Prayer in Spanish. It was followed by fairly spontaneous verses, with Latin occasionally mixed in with Quenchua [not spoken by these people]. The singing was accompanied by the rhythmic manipulation of gourds functioning as rattles, which at times were beat without singing. After about an hour of the singing, the essence of *San Pedro* was drunk by the healer, his assistant and all the patients. The cup containing the potion was tapped lightly against the stones, swords and polished sticks of the *mesa* to enhance its effects. Occasionally, a patient, the curer or his assistant left the shelter to vomit up some of the potion which was strongly emetic. More singing followed, with prayers addressed to the Virgin Mary, and to God. Occasionally, the song recited the help that the curer would give to the sick person. Each patient in his turn stood up, while the assistant removed one of the ornate swords from the table, placed it firmly in the ground between the feet of the sick man or woman and put the patient's hands upon the handle. The apprentice then took tobacco in his nose, as snuff, bending down to the ground at the feet region, the mid-section and finally the head of the patient.This was repeated with sword held firmly behind the patient. Then the sword was rubbed over every part of the patient's body, with care taken to follow the form of a cross (during this part of the ceremony, the patient stood with his arms outstretched). All of the patients were treated in this manner. The healer then spent several minutes, either privately or in front of the rest of the group, discussing the symptoms and problems of the patient. When this was completed, the songs were resumed. At one point, the assistant rose, sprinkled water in the air and cut the air with his swords to drive out "evil spirits". At another point in the ceremony, the healer rubbed together some of the many polished stones which were laid out on the cloth, causing electric sparks to fly in the darkness of the night. Finally, at dawn, the last series of songs were sung and the ceremony was over.

Healers maintain that, under the continuing effects of the cactus, they receive insight into the nature of the illness affecting their patients. The visions received, stimulated by the striking of herbal stones, are a source of pride to the healers and are cited also as the source for decisions as to which herbs to prescribe. Many healers utilize some object or a small guinea pig *(cuye)* in addition to the potion, which is passed over the body of the sick person symbolically to extract the illness. The rodent is subsequently killed, and its internal organs are examined to discover where the illness has afflicted the patient.

Much of the ritual used by the healer is syncretized with Roman Catholic beliefs, and, in fact, much of healing ritual is lifted intact from Catholic liturgy. Latin prayers, somewhat garbled but still familiar to the villagers and other patients in a land predominantly Catholic, are also recited. Prayers are addressed to various Catholic saints, who are supplicated to intercede on behalf of a sick person.

From Marlene Dobkin of California State College at Los Angeles "*Trichocereus pachanoï*—A Mescaline Cactus Used in Folk Healing in Peru." *Economic Botany.* Lancaster, Pa. XII #2 (April-June 1968) pp. 191-94. (Reprinted by permission)

Convolvulaceae or Morning-glory Family

Ololiuhqui *(Rivea corymbosa)*

BERNARDINO SAHAGUN

There is an herb which is called *coatl xoxouhqui.* It produces a seed called *ololiuhqui,* which intoxicates and causes madness. It is administered in potions to harm those who are the objects of hatred. Those who eat it have visions of terrible things. Sorcerers or persons who wish to injure some one administer it in food or drink. The herb has medicinal properties. As a remedy for gout, its seeds are ground up and applied to the affected parts.

From Fray Bernardino Sahagún (1499-1590) *Historia General de las Cosas de Nueva España.* (Translation: General History of the Things of New Spain, i.e., Mexico.) 5 volumes. Pedro Robredo. México. D.F., (1938) (Or any other edition) Book XI, ch. vii.

FRANCISCO HERNANDEZ

The *ololiuhqui,* which some call *coaxihuitl,* or 'snakeplant', is a twining herb with thin, green, cordate leaves, slender, green, terete stems; and long, white flowers. The seed is round and very much like coriander, whence the name of the plant. The roots are fibrous and slender. The plant is hot in the fourth degree. It cures syphilis and mitigates pain which is caused by chills. It relieves flatulency and removes tumours. If powdered and mixed with a little resin, it banishes chills and stimulates and aids in a remarkable degree in cases of dislocations, fractures and pelvic troubles of women. Only the seed is of medicinal use. If pulverized or taken in a decoction or used as a poultice on the head or forehead with milk and chile,

Figure 2.
Rivea corymbosa. Illustration courtesy of R. E. Schultes.

Figure 3.
Rivea corymbosa or ololuiqui of the ancient Mexicans.
From F. Hernandez *Rerum medicarum Novae Hispaniae thesaurus, seu plantarum, animalium, mineralium mexicanorum historia.*
Rome (1651). Illustration courtesy of R. E. Schultes.

it is said to cure eye troubles. When eaten, it acts as an aphrodisiac. It has a sharp taste and is very hot. Formerly, when the priests wanted to commune with their gods and to receive a message from them, they ate this plant to induce a delirium. A thousand visions and satanic hallucinations appeared to them. In its manner of action, this plant can be compared with *Solanum maniacum* of Dioscorides. It grows in warm places in the fields.

From Dr. Francisco Hernandez (1517/18-1587). *Nova Plantarum, Animalium et Mineralium Mexicanorum Historia.* (Transl. New history of Mexican plants, animals and minerals) (Added title-page: *Rerum Medicarum Novae-Hispaniae Thesaurus* (Transl.: Thesaurus of medicinal things of New Spain.) Vitalis Mascardi. Rome. (1651) Book V, ch. xiv.

JOSÉ DE ACOSTA

"Of the abominable vnction which the Mexicaine priests and other Nations vsed, and of their witchcraftes."

God appoynted in the auntient Lawe the manner how they should consecrate *Aarons* person and the other Priests, and in the Lawe of the Gospel wee have likewise the holy creame and vnction which they vse when they consecrate the Priestes of Christ. There was likewise in the auntient Lawe a sweete composition, which God defend should be employed in anie other thing then in the divine service. The Divel hath sought to counterfet all these things after his manner as hee hath accustomed, having to this end invented things so fowle and filthie, whereby they discover wel who is the Author. The priests of the idolles in *Mexico* were annoynted in this sort, they annointed the body from the foote to the head, and all the haire likewise, which hung like tresses, or a horse mane, for that they applyed this vnction wet and moyst. Their haire grew so as in time it hung downe to their hammes, so heavily that it was troublesome for them to beare it, for they did never cut it untill they died, or that they were dispensed with for their great age, or being employed in governments or some honorable charge in the commonwealth. They carried their haire in tresses, of sixe fingers breadth, which they died blacke with the fume of sapine, or firre trees, or rosine; for in all Antiquities it hath bin an offring they made vnto their idolls, and for this cause it was much esteemed and reverenced. They were alwayes died with this tincture from the foote to the head, so as they were like vnto shining Negroes, and that was their ordinary vnction: yet, whenas they went to sacrifice and give incense in the mountaines, or on the tops thereof, or in any darke and obscure caves where their idolles were, they vsed an other kinde of vnction very different, doing certaine ceremonies to take away feare, and to give them courage. This vnction was made with diverse little venomous beastes, as spiders, scorpions, palmers, salamanders, and vipers, the which the boyes in the Colledges tooke and gathered together, wherein they were so expert, as they were alwayes furnished when the Priestes called for them. The chiefe care of these boyes was to hunt after these beasts; if they went any other way and by chaunce met with any of these beasts they stayed to take them, with as great paine as if their lives depended thereon. By the reason whereof the Indians commonly feared not these venomous beasts, making no more accompt than if they were not so, having beene all bred in this exercise. To make an ointment of these beastes they took them all together, and burnt them vpon the harth of the Temple, which was before the Altare, vntill they were consumed to ashes; then did they put them in morters with much Tobacco or *Petum* (being an hearbe that Nation vseth much to benumme the flesh that they may not feele their travell), with the which they mingle the ashes, making them loose their force; they did likewise mingle with these ashes scorpions spiders and palmers alive, mingling all together; then did they put to it a certaine seede being grownd, which they call *Ololuchqui,* whereof the Indians make a drinke to see visions for that the vertue of this hearbe is to deprive man of sence They did likewise grinde with these ashes blacke and hairie wormes, whose haire only is venomous, all which they mingled together with blacke, or the fume of rosine, putting it in smal pots which they set before their god, saying it was his meate And therefore, they called it a divine meate. By means of thi oyntment they became witches, and did see and speake with the Divell. The priestes being slubbered with this oyntment lost all feare, putting on a spirit of cruelty. By reason whereof they did very boldely kill men in their sacrifices, going all alone in the night to the mountaines and into obscure caves, contemning all wilde beasts, and holding it for certayne and approved that both lions, tigres, serpents, and other furious beasts which breede in the mountaines and forrests fledde from them, by the vertue of this *Petum* of their god.

And in trueth, though this *Petum* had no power to make them flie, yet was the Divelle's picture sufficient whereinto they were transformed. This *Petum* did also serve to cure the sicke and for children, and therefore all called it the Divine Physicke; and so they came from all partes to the superiors and priests, as to their saviors, that they might apply this divine physicke, wherewith they anoynted those parts that were grieved. They said that they felt heereby a notable ease, which might be, for that Tobacco and *Ololuchqui* have this propertie of themselves, to benumme the flesh, being applied in manner of an emplaister, which must be by a stronger reason being mingled with poysons; and for that it did appease and benumme the paine, they helde it for an effect of health, and a divine virtue. And therefore ranne they to these priests as to holy men, who kept the blind and ignorant in this error, perswading them what they pleased, and making them runne after their inventions and divellish ceremonies, their authority being such, as their wordes were sufficient to induce beliefe as an article of their faith. And hus made they a thousand superstitions among the vulgar people, in their manner of offering incense, in cutting their haire, tying small flowers about their necks, and strings with small bones of snakes, commanding them to bathe at a certain ime; and that they should watch all night at the harth lest he fire should die, that they should eate no other bread but hat which had bin offered to their gods, that they should vpon ny occasion repaire vnto their witches, who with certaine raines tolde fortunes, and divined, looking into keelers and pailes full of water. The sorcerers and ministers of the divell sed much to besmere themselves. There were an infinite number of these witches, divines, enchanters, and other false prophets. There remaines yet at this day of this infection, althogh they be secret, not daring publikely to exercise their sacrileges, divelish ceremonies, and superstitions, but their abuses and wickednes are discovered more at large and particularly in the confessions made by the Prelates of *Peru.*

From José de Acosta (c. 1539-1600) *The Naturall and Morall Historie of the East and West Indies* . . . Printed by Val. Sims for Edward Blount and William Aspley. London. (1604) Book V, ch. 26. (First Spanish edition: 1590).

HERNANDO RUIZ de ALARCON

. . .Those things which I have mentioned [i.e. springs, rivers, mountains, *ololiuhqui,* etc.] the Indians believe to be deities and worship them. *Ololiuhqui* is a lentil-like seed produced by a native climbing plant. If drunk, the seed deprives them of their reason, for it is very potent. Through this potion the natives communicate with the demon, who talks to them usually when they are deprived of their senses. The demon is thus able to deceive them with various visions. They attribute the visions to the deity which they say resides in these seeds, known as *ololiuhqui* or *cuexpalli.* It is remarkable how much faith these natives have in this seed, for, when they drink it, they consult it as an oracle to learn many things that they wish to know, especially those which are beyond the power of the human mind to penetrate, as for example, to learn the cause of an illness which they attribute to witchcraft. . . . They wish to know this or find out about other things such as stolen articles, future aggressors. They consult this seed through the medium of their deceiving doctors, some of whom practice *ololiuhqui*-drinking as a profession. If a doctor who does not practice *ololiuhqui*-drinking wishes to free a patient of some trouble, he advises the patient himself to partake of the seeds. For this the patient must pay as though his doctor has taken the drink and were a *payni* [an ololiuhqui-doctor]. The doctor fixes the day and hour when the drink must be taken and names the reason for the patient's drinking it. Finally, the one drinking the *ololiuhqui* . . . must seclude himself in a room alone, mostly the doctor's parlor. No one must enter the room during the time of the divination, that is, during the time the consulting person is out of his mind. He who is consulting the seeds believes that the *ololiuhqui,* . . . is revealing what he wants to know. The delirium passed, the *payni* comes out of seclusion reciting a thousand fabrications, among which there may be a few truths. Thus in every way, the doctor keeps his patient deceived . . .

It happens that he who drinks *ololiuhqui* to excess loses his mind because of the great potency of the seed. As a consequence of indulgence the person's senses are distorted, and he who uses the narcotic speaks whatever words come into his warped mind. . . . Perhaps he condemns the innocent, perhaps he exposes the guilty, or perhaps his words are uttered in such confusion that their meaning cannot be understood . . . These unfortunate people believe the utterances, attributing everything to the deity of the *ololiuhqui* . . . For this reason they venerate and fear these plants [*ololiuhqui* and *peyote*] so much that they do all in their power so that the use of the plants does not come to the attention of the ecclesiastical authorities.

It remains to describe the manner of using *ololiuhqui,* the purpose for which it is taken, and the great inconveniences which follow its use. . . . Chronic illness, kinds which the *curanderos* have pronounced incurable with ordinary medicines, are attributed to witchcraft and . . . according to their belief, cannot be cured if the person who has cast the spell does not break it. This is the usual way in which the doctors make profits and wreak much injury with the satanic suspicions surrounding *ololiuhqui* . . . The doctor immediately attributes the illness to witchery . . . In order to aid the doctor, the patient relates his superstitions. The deceiving doctor immediately orders the use of *ololiuhqui,* and the patient follows the doctor's words as though they were the words of a prophet or of an oracle.

From B. Hernando Ruiz de Alarcon (17th century) "Tratado de las Supersticiones y Costumbres Gentilicas que oy Viven entre los Indios Naturales desta Nueva España." (Translation: Treatise on the pagan superstitions and customs today prevailing among the Indians of Mexico.) *Anales del Museo Nacional de México* Imprenta Museo Nacional. Mexico City. 6 (1898) ch. ii, vi & vii.

Tlitliltzen *(Ipomoea violacea L.)*

GORDON R. WASSON

[Below we give an account of a Zapotec *curandera* who officiated in a ceremony invoking the divine power of the morning glory seed:]

"First, the person who is to take the seeds must solemnly commit himself to take them, and to go out and cut the branches with the seed. There must also be a vow to the Virgin in favor of the sick person, so that the seed will take effect with him. If there is no such vow, there will be no effect. The sick person must seek out a child of seven or eight years, a little girl if the patient is a man, a little boy if the patient is a woman. The child should be freshly bathed and in clean clothes, all fresh and clean. The seed is then measured out, the amount that fills the cup of the hand, or about a thimble full. The time should be Friday, but at night, about eight or nine o'clock, and there must be no noise, no noise at all. As for grinding the seed, in the beginning you say, 'In the name of God and of the Virgencita, be gracious and grant the remedy, and tell us, Virgencita, what is wrong with the patient. Our hopes are in thee.' To strain the ground seed, you should use a clean cloth —a new cloth, if possible. When giving the drink to the patient, you must say three Pater Nosters and three Ave Marias. A child must carry the bowl in his hands, along with a censer. After having drunk the liquor, the patient lies down. The bowl with the censer is placed underneath, at the head of the bed. The child must remain with the other person, waiting to take care of the patient and to hear what he will say. If there is improvement, then the patient does not get up; he remains in bed. If there is no improvement, the patient gets up and lies down again in front of the altar. He stays there a while, and then rises and goes to bed again, and he should not talk until the next day. And so everything is revealed. You are told whether the trouble is an act of malice or whether it is illness."

From R. Gordon Wasson "Notes on the Present Status of *Ololiuhqui* And Other Hallucinogens of Mexico." *Botanical Museum Leaflets.* Harvard University, Cambridge, Mass. V. 20 #6 (November 22, 1963), p. 182. (Reprinted by permission.)

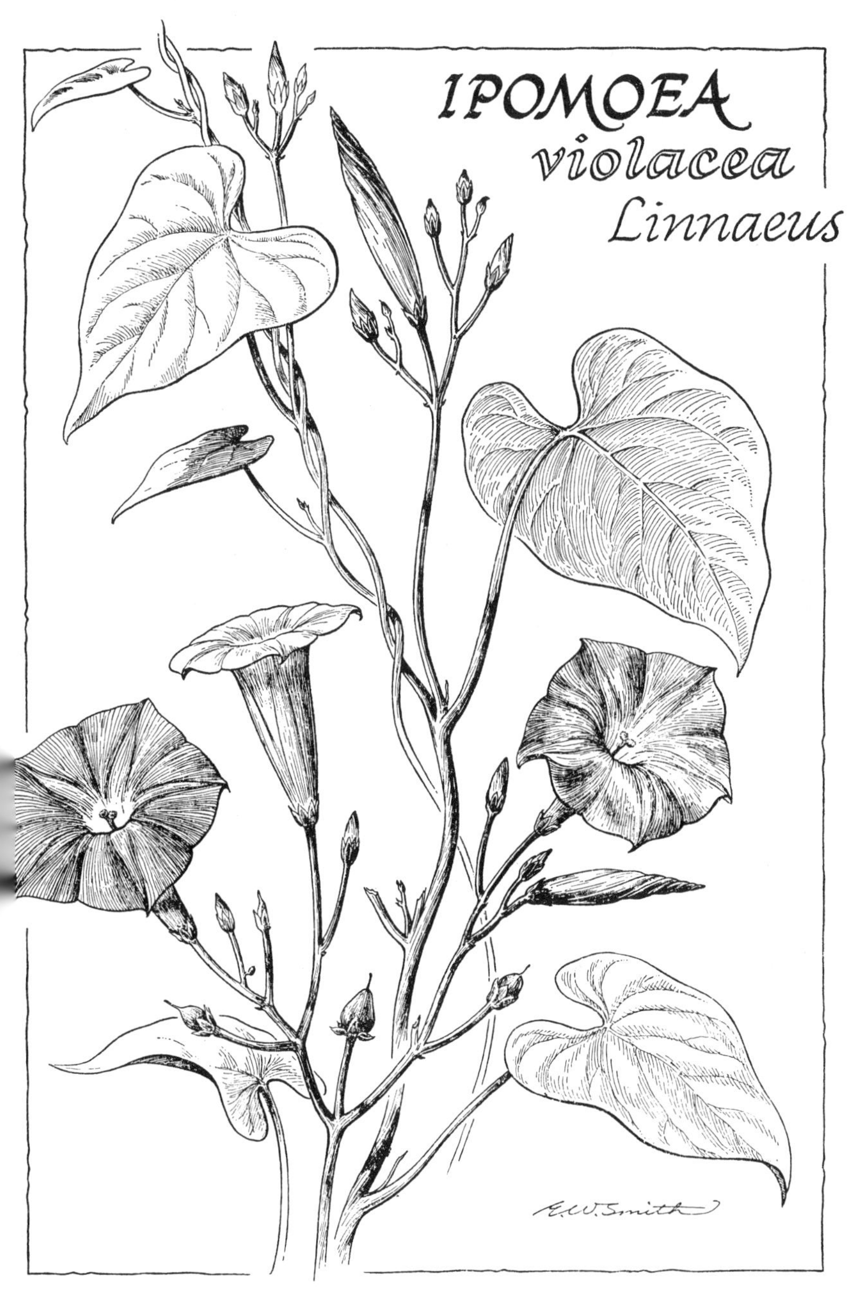

Figure 4.
Ipomoea violacea. Illustration courtesy of R. E. Schultes.

Erythroxylaceae or Coca Family

Coca *(Erythroxylon* coca, LAM.)

LOUIS LEWIN

According to an Indian legend narrated by Garcilase de la Vega the children of the sun had presented man with the *coca* leaf after the formation of the empire of the Incas, to "satisfy the hungry, provide the weary and fainting with new vigour and cause the unhappy to forget their miseries." It is probable, however, that the Indians already cultivated the plant before they formed a federation and the Incas invented the story of its divine origin in order to reserve it to themselves. They made of it a royal emblem; the queen called herself Mama Cuca and the priests assisted in upholding the divine honours of the plant by using it in various religious ceremonies. The idols of the time as a sign of divinity were represented with one cheek stuffed with coca leaves. Its use gradually extended to the people, and it was not only applied for supernatural purposes, but for the very worldly object of allowing the plant to act on the organism. Time has changed nothing in this state of affairs, except that the desire for pleasurable sensations now forms the principal motive for the use of the leaves in South America and of cocaine, their derivate, in the rest of the world.

From Louis Lewin (1850-1929) *Phantastica. Narcotic and Stimulating Drugs. Their Use and Abuse.* Foreword by Bo Holmstedt. Routledge Kegan Paul Ltd. London (1931) pp. 76-77. (Reprinted by permission)

NICHOLAS MONARDES

"Of the Coca"

I was desirous to see that hearbe so celebrated of the Indians, so many yeres past, which they call the *Coca,* which they sow and till with muche care, and diligence, because they vse it for their pleasures, which we will speake of. . . .

When they will make themselves drunke, and bee out of iudgement, they mingle with the *Coca* the leaves of the *Tabaco* and chewe them altogether, and goe as if they were out of their wittes, or as if they were drunke, which is a thing that dooth give them great contentment to be in that sort. Surely it is a thing of great consideration, to see howe desirous the Indians are to be deprived of their wittes, and to bee without vnderstanding, seeing that they vse the *Coca* with the *Tabaco,* and al to this end that they would by without vnderstanding, and have their wittes taken from them, as wee sayde in the seconde parte, when wee treated of the *Tabaco.*

From (Nicholas Monardes), Physician of Seville, c. 1512-1588. *Ioyfull Newes out of the New Found Worlde . . .* Englished by John Frampton, Marchant. Printed by E. Allde, by the assigne of B. Norton. (1596) Fol. pp. 101, 102. (First Spanish edition: Sevilla. (1569-74). First English edition: London. (1577-78)

PEDRO DE CIEZA DE LEON

"How the Indians carry Herbs or Roots in their Mouths, and Concerning the Herb Called *Coca,* which they Raise in many Parts of this Kingdom."

In all parts of the Indies through which I have travelled I have observed that the natives take great delight in having herbs or roots in their mouths. Thus, in the district of the city of Antioquia, some of the people go about with a small leaf in their mouths, and in the province of Arma they chew another leaf. In the district of Quinbaya and Anzerma they cut small twigs from a young green tree, which they rub against their teeth without ceasing. In most of the villages subject to the cities of Cali and Popayan they go about with small *coca* leaves in their mouths, to which they apply a mixture which they carry in a calabash, made from a certain earth-like lime. Throughout Peru the Indians carry the *coca* in their mouths, and from morning until they lie down to sleep, they never take it out. When I asked some of these Indians why they carried these leaves in their mouths (which they do not eat, but merely hold between their teeth), they replied that it prevents them from feeling hungry, and gives them great vigour and strength.

I believe that it has some such effect, although, perhaps, it is a custom only suited for people like these Indians. They sow this *coca* in the forests of the Andes, from Guamanga to the town of Plata. . . .

From Pedro de Cieza de Leon (1518-cc.1560) *The Travels of . . . A.D. 1532-50 Contained in the First Part of His Chronicle of Peru.* Translated and Edited by Clemens R. Markham. Printed for the Hakluyt Society. London. (1864) (Hakluyt Society) ch. xcvi. (First Spanish edition: 1553).

ABRAHAM COWLEY

. . . Of all the Plants that any soil does bear,
This tree in fruits the richest does appear,
It bears the best, and bears'em all the year,
Ev'n now with fruit 't is stor'd - . . .
Behold how thick with leaves it is beset;
Each leaf is fruit, and such substantial fare,
No fruit beside to rival it will dare.
Mov'd with his country's coming fate (whose soil
Must for her treasures be expos'd to spoil)
Our Varicocha first his *Coca* sent,
Endow'd with leaves of wondrous nourishment,
Whose juice suck'd in, and to the stomach ta'en,
Long Hunger and long labour can sustain;
From which our faint and weary Bodies find
More succour, more they cheer the drooping mind,
That can your Bacchus and Ceres join'd.
Three leaves supply for six days' march afford;
The Quitoita with this provision flor'd,
Can pass the vast and cloudy Andes o'er,
The dreadful Andes plac'd twixt Winter's store
Of winds, rains, snow, and that more humble earth
That gives the small but valiant *Coca* birth
This champion, that makes warlike Venus mirth. . . .

From Abraham Cowley (1618-1667) *Poetical Works* Four volumes. At the Apollo Press by the Martins. Edinburg, (1784) III, pp. 232-33. (Book V "The Trees, Translated by N. Tate.") (First published in Latin in 1662).

ARCHIBALD SMITH

The *coca* leaf is to the Indian of the interior a necessity of life, which he uses from time to time, to renovate his energy for renewed muscular exertion; and in the intervals of labour he often sits down to *chaccha* or to refresh himself by masticating *coca* seasoned with quick-lime, which he always carries about his person in a little gourd. The lime is used in very small quantity at a time, but in a pulverulent and escharotic state. According to the Indian it counteracts the natural tendency of the *coca* to give rise to visceral obstructions. Used in moderate quantity, the *coca,* when fresh and good, increases nervous energy, removes drowsiness, enlivens the spirits, and enables the Indian to bear cold, wet, great bodily exertion, and even want of food, to a surprising degree, with apparent ease and impunity. Taken to excess, it is said to occasion tremors in the limbs, and what is worse, a gloomy sort of mania. But such dire effects must be of rare occurrence; since, living for years on the borders of the Montaña, and in constant intercourse with persons accustomed to frequent the *coca* plantations, and with Indian yanacones or labourers, all of whom, whether old or young, masticated this favourite leaf, we never had an opportunity of witnessing a single instance in which the *coca*-chewer was affected with mania or tremor.

From Archibald Smith, M.D. *Peru as it is.* Two volumes. Richard Bentley. London. (1839) II, p. 162.

JOHANN JAKOB VON TSCHUDI

. . . The *coca (Erythroxylon coca,* LAM.) is a shrub about six feet in height, with bright green leaves and white blossoms. The latter are succeeded by small scarlet berries. . . . When the leaves are ripe, that is to say, when on being bent they crack or break off, the gathering commences. . . . After being gathered the leaves are spread out on coarse woolen cloths and dried in the sun. . . .

The Indians masticate the *coca.* Each individual carries a leathern pouch, called the *huallqui,* or the *chuspa,* and a small flask gourd, called the *ishcupuru.* The pouch contains a supply of coca leaves, and the gourd is filled with pulverized unslaked lime. Usually four times, but never less than three times a day, the Indian suspends his labor, for the purpose of masticating *coca.* This operation (which is termed *chacchar* or *acullicar*) is performed in the following manner: some of the *coca* leaves, the stalks having been carefully picked off, are masticated until they form a small ball, or as it is called an *acullico.* A thin slip of damp wood is then thrust into the *ishcupuru,* or gourd, and when drawn out some portion of the powdered lime adheres to it. The *acullico,* or ball of masticated *coca* leaves, is, whilst still lying in the mouth, punctured with this slip of wood, until the lime mixing with it, gives it a proper relish, and the abundant flow of saliva thus excited is partly expectorated and partly swallowed. When the ball ceases to emit juice, it is thrown away, and a new one is formed by the mastication of a fresh mouthfull of *coca* leaves. . . .

. . . The application of the unslaked lime demands some precaution, for if it comes in direct contact with the lips and gums, it causes a very painful burning. During a fatiguing ride across the level heights, where, owing to the cold wind, I experienced a difficulty of respiration, my Arriero recommended me to chew *coca,* assuring me that I would experience great relief from so doing. He lent me his *huallqui,* but owing to my awkward manner of using it, I cauterized my lips so severely that I did not venture on a second experiment.

The flavor of *coca* is not unpleasant. It is slightly bitter, aromatic, and similar to the worst kind of green tea. When mixed with the ashes of the musa root it is somewhat piquant, and more pleasant to European palates than it is without that addition. The smell of the fresh dried leaves in a mass is almost overpowering; but this smell entirely goes when they are packed in sacks. All who masticate *coca* have a very bad breath, pale lips and gums, greenish and stumpy teeth, and an ugly black mark at the angles of the mouth. An inveterate *coquero,* or *coca* chewer, is known at the first glance. His unsteady gait, his yellow-colored skin, his dim and sunken eyes encircled by a purple ring, his quivering lips and his general apathy, all bear evidence of the baneful effects of the *coca* juice when taken in excess. All the mountain Indians are addicted more or less to the practice of masticating *coca.* Each man consumes, on the average, between an ounce and an ounce and a half per day, and on festival days about double that quantity. The owners of mines and plantations allow their laborers to suspend their work three times a day for the *chacchar,* which usually occupies upwards of a quarter of an hour; and after that they smoke a paper cigar, which they allege crowns the zest of the *coca* mastication. He who indulges for a time in the use of *coca* finds it difficult, indeed almost impossible, to relinquish it. . . .

The operation of the *coca* is similar to that of narcotics administered in small doses. Its effects may be compared to those produced by the thorn-apple rather than to those arising from opium. . . . In the inveterate *coquero* similar symptoms [as resulting from drinking a decoction of datura] are observable, but in a mitigated degree. I may mention one circumstance attending the use of *coca,* which appears hitherto to have escaped notice: it is, that after the mastification of a great quantity of *coca* the eye seems unable to bear light, and there is a marked distension of the pupil. I have also observed this peculiarity of the eye in one who had drunk a strong extract of the infusion of *coca* leaves. In the effects consequent on the use of opium and *coca* there is this distinction, that *coca*, when taken even in the utmost excess, never causes a total alienation of the mental powers or induces sleep; but like opium, it excites the sensibility of the brain, and the repeated excitement, occasioned by its intemperate use after a series of years, wears out mental vigor and activity.

It is a well known fact, confirmed by long observation and experience, that the Indians who regularly masticate *coca* require but little food, and, nevertheless, go through excessive labor with apparent ease. They, therefore ascribe the most extraordinary qualities to the *coca*, and even believe that it might be made entirely a substitute for food. . . . Of the great power of the Indians in enduring fatigue with no other sustenance than *coca,* I may here mention an example. A Cholo of Huari, . . . was employed by me in very laborious digging. During the whole time he was in my service, viz., five days and nights, he never tasted any food, and took only two hours' sleep nightly. But at intervals of two and a half or three hours, he regularly masticated about half an ounce of *coca* leaves, and he kept an *acullico* continually in his mouth. I was constantly beside him, and therefore I had the opportunity of closely observing him. The work for which I engaged him being finished, he accompanied me on a two days' journey of twenty-three leagues across the level heights. Though on foot, he kept up with the pace of my mule, and halted only for the *chacchar.* . . .

The Indians maintain that *coca* is the best preventive of that difficulty of respiration felt in the rapid ascent of the Cordillera and the Puna. Of this fact I was fully convinced by my own personal experience. I speak here, not of the mastication of the leaves, but of the decoction taken as a beverage. When I was in the Puna, at the height of 14,000 feet above the level of the sea, I drank, always before going out to hunt, a strong infusion of coca leaves. I could then during the whole day climb the heights and follow the swift-footed wild animals without experiencing any greater difficulty of breathing than I should have felt in similar rapid movement on the coast. Moreover, I did not suffer from the symptoms of cerebral excitement or uneasiness which other travellers have observed. . . .

By the Peruvian Indians the *coca* plant is regarded as something sacred and mysterious, and it sustained an important part in the religion of the Incas. In all ceremonies, whether religious or warlike, it was introduced, for producing smoke at the great offerings, or as the sacrifice itself. During divine worship the priests chewed *coca* leaves, and unless they were supplied with them, it was believed that the favor of the gods could not be propitiated. It was also deemed necessary that the supplicator for divine grace should approach the priests with an *acullico* in his mouth. It was believed that any business undertaken without the benediction of *coca* leaves could not prosper; and to the shrub itself worship was rendered. During an interval of more than 300 years Christianity has not been able to subdue the deep rooted idolatry; for everywhere we find traces of belief in the mysterious power of this plant. The excavators in the mines of Cerro de Pasco throw masticated *coca* on hard veins of metal, in the belief that it softens the ore, and renders it more easy to work. The origin of this custom is easily explained, when it is recollected, that in the time of the Incas it was believed that the *Coyas,* or the deities of metals, rendered the mountains impenetrable, if they were not propitiated by the odor of *coca*. . . .

From Johann Jakob von Tschudi (1818-1889) *Travels in Peru, during the Years 1838-1842*. Wiley & Putnam. New York. (1847) ch. xv. (The original German edition was first published in 1846).

EDUARD FRIEDRICH POEPPIG

. . . It never has been possible to break the vice of a *Coquero,* as a true *coca* addict is called in Peru. Any *coquero* contends that he rather could bear being deprived of the most necessary than be without *coca. Coca* use has such an attraction that the desire for it increases with the age, no matter how unmistakable evil its consequences may be. It is surprising to see such a mysterious preference for a leaf, that, when fresh or dry, has only a slight odor, nothing balsamic, and when taken in small quantities, tastes merely like grass, or, at most, somewhat bitter. Any mystery, however, disappears, when through observation or personal experience it was established that *coca* is an exciting substance that can cause the same tension in the nervous system as opium does. . . . *Coca* is a source of highest glee for the Peruvian. Under its influence his habitual gloom disappears, because then his feeble imagination deludes him with pictures which he is not able to enjoy in his normal state. Although *coca* does not produce the same horrible feeling of over-excitement as opium does, it does put its users in a not unsimilar state, which is the more dangerous as this state lasts longer, albeit to a lesser degree. . . . A *coquero,* useless for any serious aim in life, more than the alcoholic, is a slave to his passion. For the enjoyment of *coca* he exposes himself to greater dangers than the alcoholic. As soon as a true *coquero* feels an irresistible desire to get intoxicated, he withdraws to solitary darkness, or to the woods, because the magic power of this herb can only then be fully felt, when the ordinary claims of life, or the distraction of associating with others which occupy his mind, completely cease.

. . . Quite apart from the undeniable influence chewing of the *coca* leaf has on the nervous system, its exciting property may also be seen from additional matters. A large heap of recently dried *coca* leaves, particularly if still retaining the warmth of the absorbed sun rays, gives off a strong odor, not unlike that of hay containing much sweet clover. This odor would give violent headaches to persons not used to the place and therefore the natives do not allow strangers to sleep in their neighborhood. Small quantities of *coca*, or after some months, do not emit this odor any longer and the less odor the less effective is the leaf. The true taste of *coca* is only then brought out, if it is taken together with burnt lime. . . . The excessive use of *coca* is always harmful to health. The common people even became aware of the moral harm that becomes manifest later only and hence they mistrust a passionate *coquero.* For a long time this abuse may go unpunished, and if there were no weekly opportunities for it, the *coquero* could reach an age of fifty years with feeling relatively little discomfort. The more frequently, however, he indulges in this vice, and the warmer, more humid, and thus more weakening a climate is, the earlier the pernicious effect of *coca* will manifest itself. Hence the Indians of the dry and cold regions of the Andes are more addict to *coca* than the inhabitants of the tropical forest, where, to be sure, other stimulants are also used by them.

A weakening of the digestive apparatus is the first symptom to be noticed on all *coqueros*. An increased or permanent abuse, however, entails an incurable disease, called there *opilación.* An early symptom of this is a slight discomfort, that easily may be mistaken for an indigestion, however, the disease soon frightfully gets worse. Bilious pains set in connected with a thousand miseries that develop under a tropical sky. Particularly constipation becomes so frequent and vexing that for this prevalence the disease was named after. Jaundice develops, and gradually then the signs of destruction in the nervous system become more clearly visible; headaches and many similar ailments set in. The sick person becomes weak and is almost unable to ingest food and rapidly emaciates. Often a kind of greenish hue on his complexion is noticeable. The bilious color turns livid . . . Incurable insomnia follows, suffered even by those who do not use *coca* in excess. Then the condition of the ill-humored sick person becomes truly pitiable, for he cannot even enjoy any longer the herb to which he owes this calamity. Besides his appetite becomes highly irregular; a strong distaste for all dishes is quite suddenly followed by a voracious appetite, especially for meat, usually beyond the reach for a poor forest dweller. Common symptoms are oedemic swellings, later turning into ascites, pains in the limbs, for a short time relieved by an eruption of boils. The sick becomes extremely moody, ordinarily he is sullen, however, he would yield to a unrestrained excess of alcoholic intoxication, if an opportunity were to occur. The *coquero* thus may drag on his pitiful existence, until, totally emaciated, he will finally die. Otherwise, psychologically, the *coquero* does not suffer as much as the hard liquor drinker, except that his inclination to isolate himself gives evil directions to his thoughts. . . .

From Eduard Friedrich Poeppig (1798-1868) *Reise in Chili, Peru und auf dem Amazonenstrom während der Jahre 1827-32.* (Translation: Travels in Chile, Peru and on the Amazon River during the years 1827-32). 2 volumes. F. Fleischer. Leipzig. (1835-36) II, pp. 210-15.

WM. LEWIS HERNDON and LARDNER GIBBON

The *coca* is a great favorite of the Quichua Indian; he prizes it as the Chinaman does his opium. While the one puts to sleep, the other keeps awake. The Indian brain being excited by *coca,* he travels a long distance without feeling fatigue, while he has plenty of *coca,* he cares little for food. Therefore after a journey he is worn out. In the city of Cuzco, where the Indians masticate the best quality of *coca,* they use it to excess. Their physical condition, compared with those who live far off from the *coca* market, in a climate inhospitable, is thin, weak and sickly; less cheerful, and not so good looking. The chewers also use more brandy and less tamborine and fiddle; seldom dance or sing. Their expression of face is doleful, made hideous by green streaks of juice streaming from each corner of the mouth.

The *coca* leaf has a very bitter taste to those unaccustomed to it. The Indians chew it with a little slacked lime, which they think eases its way down, and makes it sweater.

From Wm. Lewis Herndon and Lardner Gibbon, Lieutenants United States Navy. *Exploration of the Valley of the Amazon,* . . . 2 Parts. Part II by Lt. Lardner Gibbon. A.O.P. Nicholson. Washington. (1854) II, p. 182.

GÉRARDO REICHEL-DOLMATOFF

Upon the effect of the *coca,* the Kógi emphasize in the first place that its consumption brings a certain mental clarity which one ought to take advantage of for ceremonial gatherings and any religious act in general, being conversations, personal rites, or group rites. Evidently the *coca* causes a euphoric state which lasts for a long period and is prolonged by the gradual consumption of larger and larger quantities. The individual turns into an animated speaker and says that he feels an agreeable sensation of tingling over all the body and that his memory is considerably refreshed which permits him to speak, sing and recite during the following hours. In the second place the Kógi say that *coca* appeases hunger—according to them, however, this never is the object of consuming *coca* but only an agreeable consequence, seeing that during the ceremonies or ceremonial conversations the consumption of food is prohibited and the assistants (attendants) ought to fast. . . . Another effect which is attributed to the *coca* is insomnia. Here again the Kógi see an advantage since the ceremonial conversations should be carried on at night and individuals who can speak and sing for one or several nights without sleep merit high prestige. The Kógi ideal would be to never eat anything beside *coca,* to abstain totally from sex, to never sleep, and to speak all of his life of the 'Ancients,' that is to say, to sing, to dance and to recite.

In Richard T. Martin "The Role of Coca in the History, Religion, and Meidicine of South American Indians." *Economic Botany,* Lancaster, Pa. 24 #1 (Oct./Dec. 1970) pp. 426-28. (Reprinted by permission)

LOUIS LEWIN

Towards the middle of the sixteenth century the second Council of Lima attempted to restrict the use of *coca*-leaves by the Peruvians, Chilians and Bolivians. In canon 120 the drug is described as "a useless object liable to promote the practices and superstitions of the Indians." Political, economic, social and religious reasons gave rise to this decision. It was arrived at when the use of this substance was extensive and its cultivation was at its height, and partly because *coca* had contributed, among other causes such as drudgery and malnutrition, to a deterioration of the hygienic condition of the Peruvians. The conquistadores co-operated with the proprietors of mines and plantations; they forced the natives to labour and paid them with *coca*-leaves. In the years 1560-69, the Government prohibited compulsory labour and the administration of *coca* because "the plant is only idolatry and the work of the devil, and appears to give strength only by deception of the Evil One; it possesses no virtue, but shortens the life of many Indians who at most escape from the forests with ruined health. They should therefore, not be compelled to labour and their health and lives should be preserved." All these restrictions proved of no avail, and *coca* became a State monopoly, to pass at the end of the eighteenth century into the hands of private enterprise.

From Louis Lewin (1850-1929) *Phantastica. Narcotic and Stimulating Drugs, their Use and Abuse.* Forword by Bo Holmstedt Routledge Kegan Paul Ltd. London (1931) pp. 75-76. (Reprinted by permission)

Leguminosae or Pulse Family

Cohoba *(Piptadena peregrina)*

FERNANDO COLOMBO

[The Island of Hispaniola. The Religion of the Natives As the Admiral Christopher Columbus Himself Described it.]

. . . All these [ceremonies and customs may be found described] more copiously in the following account, which by my order was given by Brother Román, versed in their [Indian] language, on everything he could gather about ther rites´and antiquities. . . .

Brother Román's Account on the Antiquities of the Indians. . . :

I, Brother Ramón, a poor anchorite of the order of San Jerome, am writing by order of the illustrious Lord Admiral, Viceroy and Governor of the islands and mainland of the Indies, what I have been able to learn and know about the beliefs and idolatry of the Indians, and the manner they serve their gods. . . .

On What Happened to the Four [Mythical] Brothers when they were Fleeing from Giaia [the creator of the sea] . . .

. . . They soon reached the door of Bassananco's house and noticed that he had *cazzabi* . . . Deminan Caracarol [one of the brothers] went ahead of the others in order to see whether he could get some *cazzabi,* or the bread the Indians eat in this land. Caracarol entered the [baker's] house and asked for some *cazzabi,* the afore said bread. Hereupon he [the baker] put his hand to his nose, and threw a *guanguaio* [a bag] on Caracarol's shoulder; this *guanguaio* was filled with *cogioba* [*cohoba*] that he had prepared for this day; this *cogioba* is a certain powder which they sometimes take as purge and for other purposes

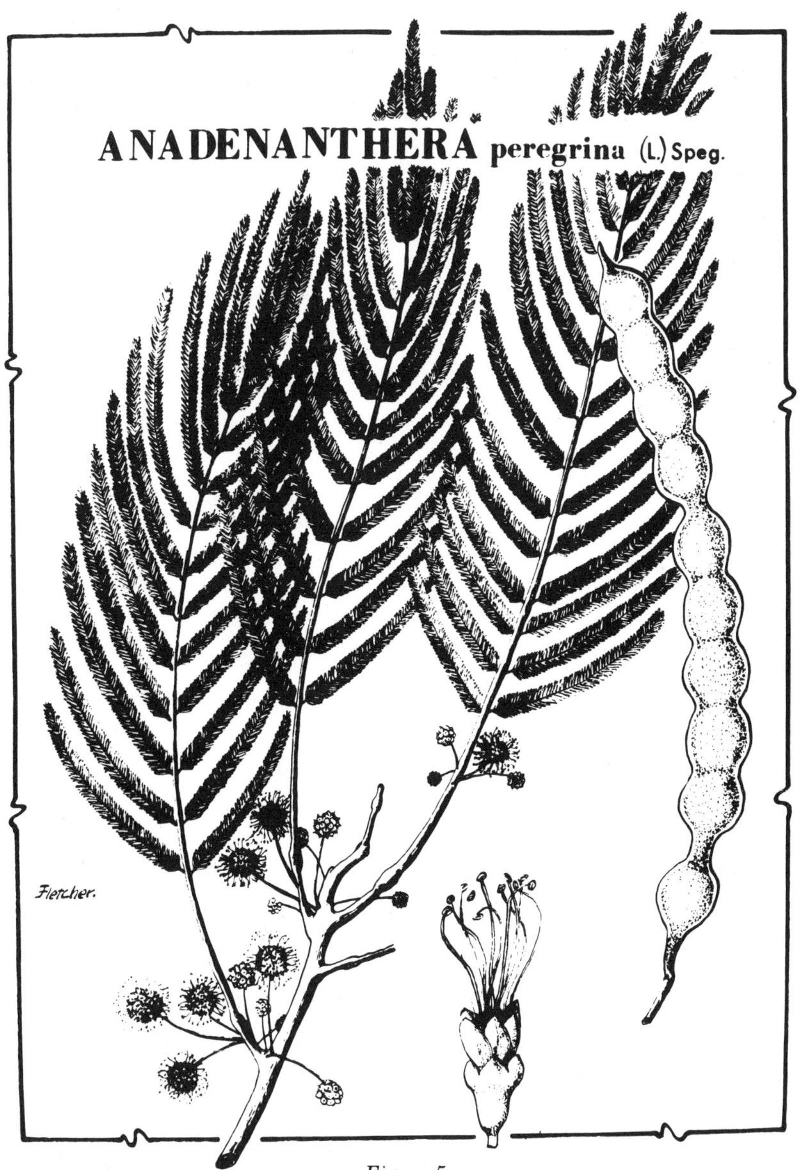

Figure 5.
Anadenanthera peregrina (syn. *Piptadena peregrina*)
Illustration courtesy of R. E. Schultes.

as we shall learn hereafter. They take this powder by means of a cane half an arm long, putting one end of it ino the nose, and the other in the powder; thus they sniff this powder up into the nose and it purges them greatly. And so he [the baker] gave them that *guangauio* instead of bread and he [Caracarol] went away very discontented because they had asked him for bread. . . .

Of How the Indian *Buhuitihus* [sorcerers or medicine men] Practice Medicine and What they Teach the People and of How they Deceive them Many Times with their Medical Cures.

. . . If an Indian is taken ill, they bring to him the *buhuitihu,* the aforesaid medicine man. This medicine man must have the same diet as the sick person and must also assume the expression of the face of this; he must also purge himself just as the sick man purges himself, that is, by taking a certain powder, called *cohoba,* snuffing it up his nose; this causes him to become so intoxicated that he does not know what he is doing. Thus the *buhuitihus* tell the people many senseless things, contending that they are speaking with the *cemies* [their wooden gods] who are telling them the cause of the illness. . . .

How They Make and Keep their Wooden and Stone *Cemies.*

[The medicine men make *cohoba* to the *cemies.*] This *cohoba* is their way of praying to the *cemies* and of asking them for fortune. . . . If they wish to know whether they will defeat their enemies, they enter a house to which only chiefs are admitted: and their lord is the first to use the *cohoba* and he also plays an instrument. While he does this, no person in the assembly may speak before the lord has finished; but having finished his prayers, the lord for a few moments remains with bowed head and with his arms resting on his knees; then he lifts his head, looks up to the sky and speaks. Hereupon all respond to him at the same time in a loud voice; and after all have spoken and thanked him, he tells them his visions he had while intoxicated by the *cohoba* that he had sniffed up his nose and which had gone to his head; he tells them that he has talked with the *cemies* and learned that they [the Indians] will carry victory over their enemies making these to flee, or, that there will be many deaths, or wars or famines, and the like, or whatever else comes to his drunken mind. Please imagine in what state his mind must be, for he seems to see houses turned upside down and persons walking with their feet sky-wards. This *cohoba* ceremony is not only offered to their stone or wooden *cemies,* but also to the bodies of their deads. . . .

From Fernando Colombo (Colón) (1488-1539) *Historie . . . Nelle quali s'ha particolare & vera relatione della vita & d'fatti dell' Amiraglio D. Christoforo Colombo, suo padre: et dello scoprimento ch'egli fecce dell' Indie Occidentali, dette Mondo Nuovo,* . . . (Translation: Histories . . . in which are given details and a true account of the life and deeds of the Admiral Don Christopher Colón, his father: And of the discovery his father made of the West Indies, called the New World, Fracesco de' Franceschi Sanese. Venice. (1571) (The Spanish original was lost after this Italian version had been made.)

BARTOLOMÉ DE LAS CASAS

On the Religion Professed by the Indians on the Hispaniola Island.

. . . We mentioned before that the Indians of this island had some, though singular, statues. It is believed that the demon talked through these statues to their priests, called *behicos,* and if he felt thus inclined, also to their lords and kings. Thus these statues were their oracles. Hence other sacrifices and ceremonies arose which they made in order to please the demon who must have taught them to do so. They proceeded in this way: they prepared some powders of certain very dry and well ground herbs of the color of ground cinnemon or privet. These finally turned tawny; the powders were placed on a dish, not flat, but rather slightly curved or deep; This dish was made of such beautiful, smooth and perfect wood, as nothing else more beautiful was ever made of gold or silver; it was almost as black and as bright as jet. Besides, they had an instrument of the same wood, as beautiful and as finished. This instrument had the form of a small flute and was entirely hollow like a flute; at two third from its bottom it opened into two tubes, such as, when extending our hand, we open our two middle fingers with the thumb withdrawn. Each of these tubes were placed into a nostril, and the front end of the flute, as it were, in the powders on the dish. Breathing the Indians sucked up the powder and snuffed it up the nose, receiving that amount of it they had determined to take. The powder thus snuffed up went to their brains, almost as if they had been drinking strong wine. In this manner the Indians became intoxicated, or almost intoxicated. The powder, the ceremonies and the procedure, they called *cohoba* (the middle syllable long) in their language. This intoxication caused them to babble confusedly, or talk like the Germans. They talked I do not know what about or with what words. This powder put them in a state to converse with the statues and oracles, or rather, with the enemy of human nature. Thus they revealed secrets, prophesized or forewarned; thus they heard or knew whether any fortune, misfortune or harm would befall them. Thus it was when the priest alone prepared himself to speak and the statues were to talk to him. When, however, all the chiefs of a village joined in performing this sacrifice or whatever it was (what they called *cohoba*), either persuaded by the *behiques* or priests, or ordered by their lords, then it was a most amusing sight. . . .

From Fray Bartolomé de las Casas (1474-1566) *Apologética Historia de las Indias.* (Translation: Apologetic history of the Indias.) Baily, Baillière é hijos, Madrid. (1909) (Or any other edition) ch. clxvi.

Yupa *(Anadenanthera peregrina)*

JOSÉ GUMILLA

. . . In their drinking-bouts, general to all Indians, the Otomacos [of the Orinoco basin], being wild and belligerent by nature, become much more enraged than other tribes. Moreover, they have another most evil habit, to wit: to intoxicate themselves through the nostrils, by using certain wicked powders which they call *yupa*. This powder totally deprives them of their reason, and ravening they grasp their weapons; if their women were not skillful in separating them and tying them up, daily much cruel havoc would be committed. This is a horrible vice. They prepare this powder of some pods of the *yupa*, hence its name. The powder proper, however, has the odor of strong tobacco. What causes the fury and intoxication is what the ingenuity of the devil makes them add to the powder. Having emptied some very large snails found on marshy grounds, they put their shells into the fire and burn them to quicklime, even whiter than snow. This lime, at equal parts, they mix with the *yupa*. Having reduced the whole to the finest powder, they obtain a mixture of diabolic strength, which is so strong, that if it were touched with the tip of his finger, the most confirmed devotee of tobacco snuff would not be able to accustom himself to it. For, if he simply put his finger which had touched the *yupa* to his nose, he would burst forth into a whirlwind of sneezes. The Saliva Indians and other tribes . . . also use the *yupa*, but being gentle, benign, and timid, they do not become maddened like our Otomacos who . . . before a battle . . . would get into a frenzy with *yupa*, injuring themselves, and smeared with blood and full of rage, would enter the battle like rabid tigers.

From Padre José Gumilla (1687?-1750) *El Orinoco, Ilustrado y Defendido* . . . (Translation: The Orinoco River, Illustrious and defended) 2 volumes. Manuel Fernandez. Madrid. (1745) I, ch. xii. (First published in 1741).

Paricá *(Piptadena peregrina)*

FRANCISCO XAVIER RIBEIRO DE SAMPAIO

At 5 p.m. we passed the mouth of the Mamiá river which discharges its waters into the Amazon in the south. . . . [Navigating down the Mamiá we reached the region inhabited by the Muras Indians.] . . . We rested at a place called *Paricátiba*, meaning a place with an abundance of paricá trees. The fruit of this tree is roasted and ground to a fine powder. This powder is commonly more valued by the Indians than tobacco. This powder the Indians use at a festival called *parassé*, named after the paricá. A big hunt without distribution is arranged in the villages. The ceremonies performed at the festival are as follows: First, the men whip one another with lashes, the length of a man's arm. These lashes are made of the hide of the manates, tapir, or stag, or, in their stead, a well twisted rope made of pita (the fibers of the agave leaves) is used. A stone or any other handy solid material is fastened to the end of the lash. With this instrument they whip one another, one man standing with his arms spread out, while the other whips him to his heart's content; then, in his turn, the whipped whips the whipper. This most cruel ceremony lasts for eight days, while the old women prepare the *paricá,* and the young women make fruit wine and tapioca cakes, called *payavarú*. The men, having finished whipping one another, enter the hut to take the *paricá*. Those who before whipped one another, now enjoy themselves together. The *paricá* is taken in this manner: each of the whipping companions holds a pipe filled with this powder, first, placing the ends into the right nostril of his companion, while the other man blows with incredibly great force and immediately again refills the pipe, repeating this operation in the left nostril of his friend. Then the latter does the same to the former. This they perform during the entire day, and in the evening they begin to drink and drink during the entire night. The power of the *paricá* together with the strength of the wine are so violent that those who have taken them fall to the ground like dead and it happens many times that some of the men die, suffocated with the *paricá*. Those, however, who wake up after the drinking bout is over, again return to the festival, continuing it for the balance of the days it is still lasting. This festival takes place every year. They celebrate it when they recruit soldiers, or when they initiate young men into virility.

From Francisco Ribeiro de Sampaio (1741-cc. 1813) *Diario da Viagem . . . da Capitania da Rio Negro no Anno de (1774-1775)* (Translation: Travel diary . . . of the district of Rio Negro during the Years 1774 and 1775). Typografia da Academia. Lisbo. (1825) Section lviii.

JOHANN BAPTIST VON SPIX & CARL FRIEDRICH PHILIP VON MARTIUS

The Muras Indians, a wandering tribe, have a very strange custom, peculiar only to them. This is the use of the *paricá* snuff. For preparing the paricá powder, they use the dried seeds of the *Parica üva,* an *Inga* species. First, the effect of the powder is exciting, then, narcotic. Once a year, when the seeds are ripe, the members of each tribe use the *paricá* for eight days. This practice is accompanied by constantly drinking intoxicating beverages, dancing and singing. For this feast the whole tribe assembles in a spacious, open house, where the women animate the men with *cujas* of the *cajiri,* abundantly supplied, or with other vegetable drinks. Then the men arrange themselves in pairs, according to their choice, and whip one another with leather straps until they draw blood. This odd whipping is not a hostile act, but, on the contrary, an act of friendship. This entire excess, as we were told, may be considered a misdirected sexual relationship. After this bloody operation, lasting for several days, is over, the pairs blow one another *paricá* powder into their nostrils with a tube made of a hollow tapir bone, one foot long. They blow with such a force and so continuously, that sometimes some men die, either suffocated with the fine dust thrust up to their frontal cavity, or, from being over-excited from the narcotic effect of the powder. With unparalleled rage the pairs fill the *paricá* from large bamboo pipes *(tabocas)* in which the powder is kept. A hollow crocodile tooth serves as measure for each insufflation. With their knees drawn close to one another, the men blow and stuff the powder into their nostrils. The effect of this is a sudden exaltation, nonsensical talking, screaming, singing, wild jumping and dancing. Having been stupefied by drinks and all kinds of dissipation, the men then fall into a beastly drunkenness. It is said that a decoction of *paricá* is also used as a clyster producing a similar, though weaker effect.

Paricá snuffing may also be found with the Mauhés Indians, but since this is a more civilized tribe, it is taken in a more refined manner.

Finally, the Miranhas use the *ypadú* powder, made of the leaves of *Erythroxylon coca,* LAM. Some travelers found this custom also with Peruvian Indians.

From Johann Baptist von Spix (1781-1826) and Carl Friedrich Philip von Martius (1794-1868) *Reise in Brasilien in den Jahren 1817 bis 1820.* (Translation: Travels in Brazil during the years from 1817 to 1820). 3 volumes. M. Lindauer. Munich. (1823-31) pp. 1074-75.

Paricá *(Piptadena* spp.)

MAUGIN DE LINCOURT, Alphonse

"Fragments of Travels from Itaibu to the Cataracts of the Tapajos, and among the Mundrucus and Maués Indians."

. . . The Indian . . . took his *paricá*. He beat, in a mortar of *sapucaia,* a piece of hard paste, which is kept in a box made of a shell; poured this pulverized powder upon a dish presented by another Indian, and with a long pencil of hair of the *tamandua bandeira,* he spread it evenly without touching it with the fingers; then taking pipes joined together, made of the quills of the *gavião real,* (royal eagle), and placing it under his nose, he snuffed up with a strong inspiration all the powder contained in the plate. His eyes started from his head; his mouth contracted; his limbs trembled. It was fearful to see him; he was obliged to sit down, or he would have fallen; he was drunk, but this intoxication lasted but five minutes; he was then gayer.

Afterwards, by many entreaties, I obtained from him his precious *paricá,* or rather one of them, for he possessed two.

At the Malocca of *Taguariti,* where I was the next day, the Tuchão, observing two young children returning from the woods laden with sarsaparilla, covered with perspiration, and overcome, as much by the burden they carried as the distance they had travelled, called them to him, beat some *paricá*, and compelled them to snuff it.

I then understood that a Tuchão Mahué had a paternal authority in his malocca, and treated all as his own children. He forced these children to take the *paricá*, convinced that by it they avoided fevers or other diseases. And, in truth, I soon saw the children leave the cabin entirely refreshed, and run playing to the brook and throw themselves in.

Several vegetable substances compose *paricá:* first, the ashes of a vine that I cannot class, not having been able to procure the flowers; second, seeds of the *acacia angico,* of the leguminous family; third, the juice of the leaves of the *abuta,* (cocculus) of the menispermes family. . . .

From Alphonse Maugin De Lincourt, French Engineer and Architect. "Fragments of Travels from Itaituba to the Cataracts of the Tapajos, and among the Mundrucus and Maués Indians." In: Lieut. Wm. Lewis Herndon (1813-1857) *Exploration of the Valley of the Amazon, Made Under Direction of the Navy Department.* 2 parts. Robert Armstrong. Washington. (1854) 1, pp. 314-15.

Niopo *(Anadenenanthera peregrina)*

RICHARD SPRUCE

Niopo Snuff and the Mode of Using it.

. . . We owe our first knowledge of *Niopo* snuff and of the tree producing it, to Humboldt and Bonpland, . . .

[In 1851] I had . . . purchased of a Brazilian trader at Manáos an apparatus for taking *niopo* snuff . . . Hed had brought it from the river Purús, where it had been used by the Catauixi Indians. My note on it (as taken from his account) is as follows:

The Catauixis use *niopo* snuff as a narcotic stimulant, precisely as the Guahibos of Venezuela, and as the Múras and other nations of the Amazon, where it is called *paricá*. For absorbing *paricá* by the nose, a bent tube is made of a bird's shank bone, cut in two, and the pieces joined by wrapping, at such an angle that one end being applied to the mouth, the other reaches the nostrils. A portion of snuff is then put into the tube and blown with great force up the nose. A clyster-pipe is made, on the same principle, of the long shank bone of the Tuyuyú *(Mycteria americana)*. The effect of *paricá*, taken as snuff, is to speedily induce a sort of intoxication, resembling in its symptoms (as described to me in this instance) that produced by the fungus *Amanita muscaria*. Taken in injection, it is a purge, more or less violent according to the dose. When the Catauixi is about to set forth on the chase, he takes a small injection of *paricá*, and administers another to his dog, the effect on both being (it is said) to clear their vision and render them more alert.

From Richard Spruce (1817-1893) *Notes of a Botanist on the Amazon & Andes* . . . In Two Volumes. Macmillan & Co., Ltd., London. (1908) II, pp. 426-29.

RICHARD SCHOMBURGK

"*Mimosa acaciodides* BENTH, or *Paricá*, or *Paricarama*, a Native Opiate."

[On the banks of the Rupunni River] grow a large number of trees of the *Mimosa acacioides* BENTH., the *paricá* or *paricarama* of the Indians. The British Guiana aborigines apply its seeds to the same purpose as the Otomacs and Guajibos of Orinoco put to use the beans of *Acacia Niopo* HUMB. BONP., and Asiatic peoples use opium. They pound the beans to a fine powder, burn it and inhale the smoke or else rub it into the eyes and ears. Either method soon puts them into a drunken and ecstatic condition lasting several hours, in its extreme degree bordering on madness, that is succeeded by a stage of great exhaustion and drowsiness.

From Richard Schomburgk (1811-1891) *Travels in British Guiana, 1804-1844)* Translated and edited . . . by Walter E. Roth. 2 volumes. "Daily Chronicle" Office. Georgetown, British Guiana. (1922-23) II, p. 204.

Yopo *(Piptadena peregrina)*

CARLOS ALMA YBARRA

. . . There are lives that are eternal nights, drunk with *yopo* and embalmed by tropical forests.

From Carlos Alama Ybarra (1900 -) *Rio Negro* Tipographía Vargas. Caracas. (1950) pp. 53-57.

OTTO ZERRIES

[When looking into the mythology and spiritual culture of the Waika on the Upper Orinoco] we discovered among other things a comprehensive system of nature spirits, the so-called *hekula* . . . As they are celestial beings, the *hekula* dwell in heaven, but they also inhabit the sphere between heaven and earth. In animal form, more rarely in human form, most *hekula* as species spirits control animals and plants on earth, living in the *ulihihama,* the forest. Through the *hekulamo,* the invocation of the medicine man, however, they come from the mountains, or walk along the sky. They travel with the wind and therefore the wind is either cheerfully greeted or anxiously warded off by the Indians, depending on whether the *hekula* reveal in it a friendly or hostile disposition. In a small form the *hekula* penetrate the chest of the medicine man, who with certain incantations can gain power over a number of them. For the *hekulamo* it is indispensable for the medicine man to take *yopo*-powder. The blowing of *yopo*-powder may perhaps be related to the moving with the wind of the *hekula* spirits. Waika woman neither snuff nor practice *hekulamo*. . . .

. . . Only after having entered his chest can the *hekula* be seen by the medicine man in his *yopo* intoxication. Overpowered by the drug and his own imagination, the medicine man often goes out of his senses. Sometimes he is seized with convulsions and the spirit which laid hold of him thus becomes directly manifest and the medicine man must be exorcized by his colleagues. At times he loses consciousness, a condition which the Waikas revealingly give the name "dying". Presumably this effect is produced by the alkaloid of *Piptadena peregrina,* a plant resembling a mimosa, which is one of the numerous ingredients of the snuff. The plant *hisioma* from which this ingredient of the *yopo*-powder is taken, is subordinate to two *hekula* spirits, the *ihamaliwa* and the *kuhidiliwa*, the lord of the sloth and of an (unknown) bird species. In primeval times these two spirits together brought down to earth the image soul *(noúdible)* of *hisioma* and thus created this plant. . . .

From Otto Zerries "Medizinmannwesen und Geisterglaube der Waika-Indianer des Oberen Orinoko". (Translation: Medicine men and belief in spirits of the Waika Indians on the Upper Orinoco). (24th Frobenius Expedition, 1952-54 in Bolivia . . .). *Völkerkundliche Forschungen Martin Heydrich zum 70. Geburtstag überreicht von Freunden und Schulern* Herausgegeben von W. Fröhlich. (Festschrift in honor of Martin Heydrich . . .). E. J. Brill. G.m.b.H. Cologne. (1960) Introduction: I & III. (Translation by permission)

Vihó *(Piptadena* spp.)

GERARDO REICHEL-DOLMATOFF

Most religious rituals [of the Desana Indians, Colombian Northeast Amazon] are the duties of the payé *(ye'é),* who serves as an intermediary between society and the supernatural forces. The payé directs the ceremonies of the life cycle, that is, "baptism", initiations, and burial rites. Besides, he is also a curer. But his central function has an essentially economic character as he is the intermediary between the hunter and the supernatural "masters" of the animals. . . . In his contacts with supernatural beings, the payé uses certain hallucinogenic drugs, such as the powder of *vihó (Piptadena),* which he sniffs in through his nose, or the drink *gahpi (Banisteriopsis caapi)* [of the Malpighiaceae or Malpighia Family], called *yayè* in Lengua Geral.

The characteristic paraphernalia of the payé consist of his gourd rattle and a long lance-shaped rattle as well as an ornament formed by a polished cylinder or white or yellowish quartz that he wears suspended from his neck. . . .

The sun created the various beings, so that they would represent him and serve as intermediaries between him and the earth. To these beings he gave the duty of caring for and protecting his creation and of promoting the fertility of life. . . .

[The Sun] . . . created *Vihó-mahsë,* the Being of *Vihó,* the hallucinogenic powder, and ordered him to serve as an intermediary so that through hallucinations people could put themselves in contacts with all the other supernatural beings. . . . the Sun gave *Vihó-mahsë* the power of being good and evil and put him in the Milky Way as the owner of sickness and witchcraft.

Then the Sun created *Vaí-mahsë,* the Master of Animals. . . .

The Sun had the *vihó* powder in his navel, but a daughter of *Vaí- mahsë* owned the *yajé* plant. She was pregnant and with the pain of childbirth she went to the beach, and, lying down, twisted in pain. An old Desana woman wanted to help her and took hold of her hand, but the daughter of *Vaí-mahsë* twisted so hard that she broke her finger, and the old woman kept it. She kept the finger in her maloca, but a young man stole it and planted it. The *yajé* plant originated from this finger. . . .

. . . the Milky Way is interpreted as an immense seminal flow that fertilizes all of the intermediate zone, or the underlying biosphere. This principle of fertilization has, however, a somewhat ambivalent character. In the first place, the Milky Way is the zone of communication where contact between terrestrial beings and supernatural beings is established. These contacts are obtained by means of hallucinogenic drugs or, at least, by means of visions induced by a state of profound concentration. The Milky Way is directly designated as the "zone of hallucinations and visions" into which the payé and other persons who take a hallucinogenic drug can penetrate and thereby pass from one cosmic level to another. This zone is dominated by

Vihó-mahsë, the divine personification of the *vihó* powder *(Piptadenia)* who, in a state of perpetual trance, travels along this celestial way observing the earth and its inhabitants. In their trance the payés rise to the Milky Way to ask *Vihó-mahsë* to serve as intermediary with the other divine personifications, *Emëkóri-mahsë* [the Being of Day], *Diroa-mahsë* [the Being of Blood], or *Vaí-mahsë* [the Master of Animals]. But on the other hand, the Milky Way is the dwelling place of the sicknesses. It can be thought of as a large, rising river in whose turbulent and foaming waters float residue and waste; these are the essences of putrefaction and, consequently, are very dangerous pathogenic factors for living beings. Putrefaction is the same as illness, and *Vihó-mahsë* can channel the current in such a way that illnesses come to contaminate the earth. . . . *Vihó-mahsë* occupies a very ambiguous position and can cause good as well as evil. . . . the Milky Way is a zone of great danger. In the dimension of hallucinations man can obtain good and find divine illumination, but, in taking undue advantage of this state, he can also cause evil. . . .

Vihó-mahsë is the divinity associated with the use of the powder of *vihó (Piptadenia)* and, by extension, with all the hallucinogenic plants. The name is derived from *vihíri,* to inhale, to absorb, and refers to fine powder that the payés sniff through the nostrils by means of a small tubular bone. The *Vihó-mahsë* (the name is used at times in the plural) are the most important intermediaries in shamanistic practices because, for any ritual action, the payé must first put himself in contact with them to ask for their assistance. They live under the waters and also in the hills where they occupy huge malocas *(vihó-mahsá vi'i),* but their true sphere of action is the Milky Way from whose heights they observe the doings of mankind. The *vihó-mahsá* are essentially amoral beings because they also serve evil persons who want to cause harm to an enemy. In this case the intermediaries come to be the direct causes of evil because they can disturb the currents of the Milky Way and send sickness to the earth.

At times, the *vihó-mahsá* seek contact with the payés and then manifest themselves in the form of black clouds gathering over a hill or a large cliff. When this phenomenon is observed, the payé puts himself into a trance and establishes contact with them to find out their intentions. In this way the payé of a neighboring tribe can transmit threats, formulate complaints, and convoke a gathering of several payés who then, in their hallucinations, converse and decide the fate of an individual. They decide about the course of sickness or about the way in which hunting or fishing should be carried out. When black clouds gather over the forest, people say: "This hill wants to do us harm," and the payé immediately reacts by taking *vihó* to find out the desires of the *vihó-mahsá*. . . .

The only ones who know the hills . . . are the payés because their role is to speak with *Vaí-mahsë* [the Master of Animals] so that he will cede some of his animals to the hunter. In a state of hallucination induced by absorbing the narcotic powder of *vihó*, and with the help of *Vihó-mahsë* who serves as an intermediary, the payé enters the hill to negotiate with *Vaí-mahsë* He does not ask for individual animals but asks for herds or a good season, and as "payment" he promises to send to the house of *Vaí-mahsë* a certain number of souls of persons who, at their death, must return to this great "storehouse" of the hills to replenish the energy of those animals the Master of Animals gives to the hunters. . . .

The office and shamanistic power are directly derived from the Sun Father who was the first payé. The Sun guarded the narcotic powder of *vihó* in his navel but, the myth says, ". . . women scratched the navel and gave the powder to their relatives". Since then, *vihó* is the main attribute of the payés . . .

It is about twenty-five years of age that a person begins to show [the qualities to become a payé.] and it is then that formal [apprenticeship] training is begun. . . . The apprenticeship lasts some six months or longer, during which the payé communicates to the apprentice the invocations, the myths referring to the creation of the Universe, . . . Also, after three or more months, the neophyte practices the use of hallucinogenic drugs, and together they sniff the powder of *vihó* and drink *yajé*. Using a long tube, the payé blows the powder into the nostrils of his pupil, communicating to him not only the snuff but also the power to have visions. . . .

[One of the Desana myths tells that] "there is a certain

snake that has a large scale in the shape of a gourd cup. The exterior is like sandpaper. A Desana, taking *vihó* snuff, threw himself into the river and found the scale. He was living on the headwaters of a small stream. He kept the scale in his bag together with his crystals and his *vihó* without giving it any importance. Then many snakes of different kinds came, snakes that he had never seen before. Then he spoke to another payé. This one told him that certainly he had something, something that belonged to a snake, and that he ought to get rid of this thing. The man threw the scale into the river, but the water was very shallow because the stream was small. At night the stream grew, and many snakes came, followed by fish. The place where he had thrown away the scale became a lagoon; it became the best fishing ground."

From Gerarde Reichel-Dolmatoff *Amazonian Cosmos. The Sexual and Religious Symbolism of the Tukano Indians.*

Vilca *(Anadenanthera colubrina)*

BERNABÉ COBO

On the Vilca

In Peru grows a tree which is called *vilca* by the Indians; it is as tall as an olive tree and has small dark-green leaves resembling those of the *guarango (Acacia tortuosa)*. It has abundant foliage and is pleasant to look at. It produces some dry capsules like carob pods. These pods, the length of the third of a vara and the width of two fingers, contain some seeds as large and thin as half a real. The skin of the pods is smooth, dark tawny and very thin. The seeds contain a yellow pulp tasting as bitter as the seeds of the *alcibar*. The Indians attach great value to these seeds because of their medicinal properties. They use them to cure certain illnesses, such as fever, hemorraghes or intermittent fevers. They take this purge in *chicha* (corn-wine) which is their ordinary beverage. The seeds have laxative power causing them to vomit and thus to evacuate their choleric and melancholic temperaments. A concoction of these taken with honey clears the chest and stomach and also has diuretic effects. The Indians also affirm that it makes their women fertile. The tree is much prized for its hard wood and hence it is used by the Indians for making many articles requiring strong wood. . . . (Book VI, ch. lxxxix).

. . . The disciples of the demon were called *umu* by the Indians who considered them to be wizards. Hence the Indians resorted to them when they wished to find out about articles lost or stolen, about future success, or about what was happening in distant places. These wizards talked to, and consulted the demon in dark places. The demon answered them in a hoarse and terrifying voice that sometimes could be heard by the rest of the Indians, though they could not understand what this voice said, nor could they see from where it came. Sure enough, these wizards made extraordinary discoveries of things lost or stolen. . . . As regards future happenings, they commonly lied, to be sure. This, however, did not harm their reputation, for then they alleged that the demon had changed his mind. Usually they invoked the demon in different manners: either, by drawing certain lines and circles on the ground and muttering the proper words; or, at other times, by entering a room which they locked from inside and where they used certain ointments whereby they got so intoxicated that they lost their senses; then, the following day, they told the people what they had asked the demon. For consulting and conversing with the demon, the wizards performed a thousand ceremonies and sacrifices. The principal ceremony was to get intoxicated with *chicha* which they had mixed with the juice of a plant called *vilca*. . . . (Book XIII, ch. xxxvi.)

From Padre Bernabé Cobo (S.J.) (1582-1657) *Historia Natural y Moral de las Indias.* (Translation: Natural and Moral History of the Indias). [Written in 1653], 4 volumes. E. Rasco. Sevilla. (1890-93) (First edition).

Vinho da Jurema *(Mimosa hostilis)*

OSVALDO GONÇALVES de LIMA

[In 1942, while visiting the Pancarú Indians of Tacaratú (Pernambuco), due to lack of time and pressure of work, we could not attend an *ajucá* festival which previously had been faithfully described by Carlos Estevão de Oliveira.]

However, we had the opportunity to observe one part of the ritual, the preparation of the *vinho da Jurema* by Serafim Joaquim dos Santos, the centenarian chief of the tribe.

The evidence and techniques he employed coincide with those described by Carlos Estevão whom we wish to quote below:

". . . Not all of the villagers are allowed to participate in the *jurema* or *ajucá* festival because of its secret nature, being essentially religious. [The festival is usually held at night time in the middle of the forest.] On my request, old Serafim was willing to hold the festival during the day and allowed me to attend it. [On the day fixed, I and two companions went to Serafim's house.] On our arrival we found that the chief was already at the place where the festival was to take place. He was just about to prepare the *ajucá*, the miraculous drink made from the jurema root. I witnessed the entire preparation. First the root is scraped, then washed in order to eliminate any dirt that might still cling to it. Then it is placed on a stone and beaten again and again with another stone in order to crush it. Afterwards the completely crushed mass is thrown into a vessel filled with water, and pressed out by hand by the person who prepares it. The water gradually turns into a reddish and frothy syrup, at which point it is ready to drink. The froth is quickly removed, thus leaving a clear liquid. Now old Serafim lighted a tubular pipe made from the *jurema* root and placed the lit end into his mouth. Then he blew on the liquid in the vessel, making the smoke form the figure of a cross . . . Thereupon an Indian, a son of the chief, placed the vessel on the ground upon two *uricuri* leaves which formed a kind of mat. Then those present, among them two old women who were well-known as singers, sat on the ground in a circle around the vessel. Presently the festival began. A religious atmosphere

Figure 6.
Mimosa hostilis. Illustration courtesy of R. E. Schultes.

filled the vault of foliage which sheltered us. With a repentant expression on their faces, their heads bent, they fixed their eyes on the ground. . . . The vessel went around, passing from hand to hand and returning to the master. At that moment, one of the old women, playing the *maracá*, began to sing. It was an invocation to Our Lady, asking Her for peace and happiness for the village. Then they chanted pagan songs of sorcery. In the songs from time to time could be heard the names of Jesus Christ, Mother of God, Our Lady, Eternal Father, and sometimes also the name of father Cicero. In a song addressed to Our Lady, one of the women singers thanked Her for my presence in the village and asked Her to give me happiness. During this time, the Indian who had placed the vessel on the leaves respectfully and solemnly offered to the rest of the people this magic drink which transports the individual to strange worlds, and allows him to enter into contact with the souls of the dead and with their protective spirits.

He who received the vessel sipped with greatest reverence some draughts of the *ajuca.* When the drink returned again to the first of the old women singers, old Maria Pastora stood up, took the vessel, lifted both her hands over her head, and, looking skywards, recited a prayer in low voice. Then she sat down, drank the *ajuca.* Having distributed the drink to all those present, the chief's son knelt down on the *uricuri* leaves and in his turn sipped some of the drink. The rest of the liquid was put into a hollow prepared for this purpose. All these scenes were accompanied by songs and *maricá* playing. When one woman singer rested, the other began to sing. Now another person started the pipes around the circle, passing from hand to hand, from mouth to mouth. Finally, all got up, men and women. Thereupon the women began again to sing and play the *maricá*. They blessed those attending, one by one, myself included, always singing. Maria Pastora blessed me, invoking God for me and praying for my happiness. The other singer, this time invoking Our Lady, blessed me and called me "the wayfarer to the villages". Thereupon Maria Pastora continued by calling all those who were nearby but not allowed to attend the *ajucá*, and she blessed them all. Finally, the two women took leave, promising their future services to the chief. Before leaving, however, Maria Pastora stood and murmured some prayer to one of the spirits protecting the village."

... We succeeded in obtaining a song used by the women during the *ajucá* libations:

"My little Ajucá
From where do you come?
I come from the world
From the world I come
I come painting
I come messing up
I come imploring
From the house of the Lord Master."

...

From Osvaldo Gonçalves de Lima "Observacões sobre o 'Vinho da Jurema' Utilizado pelos Indios Pancarú de Tacaratú (Pernambuco)." (Translation: Observations regarding 'Vinho da Jurema' used by the Pancarú Indians of Tacaratú, Pernambuco).Instituto de Pesquisas Agronomicas. *Arquivos.* Recife, Pernambuco Brazil. (Archives of the Institute for agronomic research). 4 (1946) pp. 46-50. (Reprinted by permission)

Yurema *(Mimosa hostilis)*

ROBERT H. LOWIE

In 1938 [Curt] Nimuendajú [of Belém do Pará] (mss.) gleaned a few facts about the ancient *yurema* cult. An old master of ceremonies, wielding a dance rattle decorated with feather mosaic, would serve a bowlful of the infusion made from *yurema* roots to all celebrants, who would then see glorious visions of the spirit land with flowers and birds. They might catch a glimpse of the clashing rocks that destroy souls of the dead journeying to their goal, or see the Thunderbird shooting lightning from a huge tuft on his head and producing claps of thunder by running about.

From Robert H. Lowie (1883-1957) "The Cariri [Eastern Brazil]" *Handbook of South American Indians.* U.S. Government Printing Office, Washington, D.C., (1946) I, p. 559.

Malpighiaceae or Malpighia Family

Caapi, Aya-huasca, Yajé *(Banisteriopsis* spp.)

MANUEL VILLAVICENCIO

We shall not pass over in silence something that seems most remarkable to us. This is a liana which the Zaparos, Santa Marias, Mazanes and Angutéros use effectfully for divining, foreseeing, or for answering with authority in difficult cases, such as giving the proper reply to ambassadors of other tribes in matters concerning going to war. They also use this magic drink to discover the enemies' planes and to take the convenient measures for attack and defence; or if a member of their family is taken ill, to discover by which witchcraft he was brought into this state . . . finally to assure themselves of the love of their wives. The procedure is the following: they prepare a light dedoction of the liana called *aya-huasca* (the liana of death or souls). The Indian who must give the answers and make the plans takes this brew, but mostly all the assembled Indians drink it. It is of course a narcotic drink and after a short time it begins to produce the strangest phenomena. It seems to affect the nervous system, for all senses become sharpened, all mental faculties awaked. . . . As for me, when I took the *aya-huasca* drink, I first felt my head to swim, then to be taken on a trip through the air, seeing the most charming landscapes, large cities, lofty towers, beautiful parks and other delightful things. But suddenly I found myself deserted in a forest, attacked by wild beasts of prey against which I tried to defend myself; finally I had a strong sensation of sleep, accompanied by headaches, and occasionally a general *malaise*. . . .

From Manuel Villavicencio, Medical Doctor and Member of Various Scientific Academies. *Geografia de la República del Ecuador.* Robert Craighead. New York. (1858) pp. 371-73.

Figure 7.
Banisteriopsis caapi. Illustration courtesy of R. E. Schultes.

RICHARD SPRUCE

[The *Banisteriopsis Caapi* is a woody twiner growing] on the river Uaupés, the Içanna, and other upper tributaries of the Rio Negro, where it is commonly planted in the roças or mandiocca-plots; also at the cataracts of the Orinoco, and its tributaries, from the Meta upwards; and on the Napo and Pastasa and their affluents, about the eastern foot of the Equatorial Andes. Native names: *caapi,* in Brazil and Venezuela; *Cadána,* by the Tucáno Indians on the Uaupés; *aya-huasca* (i.e. Dead man's vine) in Ecuador.

The lower part of the stem is the part used. A quantity of this is beaten in a mortar, with water, and sometimes with the addition of a small portion of the slender roots of the *caapi-pinîma (Haemadictyon amazonicum).* When sufficiently triturated, it is passed through a sieve, which separates the woody fibre, and to the residue enough water is added to render it drinkable. Thus prepared, its colour is brownish-green, and its taste bitter and disagreeable.

"The Use and Effects of *Caapi*": In November 1852 I was present, by special invitation, at a *Dabocuri* or Feast of Gifts, held in a *mallóca* or village house . . . above the first falls of the Uaupés. . . . We reached the *mallóca* at nightfall, just as the *botútos* or sacred trumpets began to boom lugubriously within the margin of the forest skirting the wide space kept open and clear of weeds around the *mallóca.* At that sound every female outside makes a rush into the house, before the *botútos* emerge on the open; for to merely see one of them would be to her a sentence of death. We found about 300 people assembled, and the dances at once commenced. . . .

In the course of the night, the young men partook of *caapi* five or six times, in the intervals between the dances; but only a few of them at a time, and very few drank of it twice. The cupbearer—who must be a man, for no woman can touch or taste *caapi*—starts at a short run from the opposite end of the house, with a small calabash containing about a teacupful of *caapi* in each hand, muttering "Mo-mo-mo-mo-mo" as he runs, and gradually sinking down until at last his chin nearly touches his knees, when he reaches out one of his cups to the man who stands ready to receive it, and when that is drunk off, then the other cup.

In two minutes or less after drinking it, its effects begin to be apparent. The Indian turns deadly pale, trembles in every limb, and horror is in his aspect. Suddenly contrary symptoms succeed: he bursts into a perspiration, and seems possessed with reckless fury, seizes whatever arms are at hand, his *murucú,* bows and arrows, or cutlass, and rushes to the doorway, where he inflicts violent blows on the ground or the doorposts, calling out all the while, "Thus would I do to mine enemy (naming him by his name) were this he!" In about ten minutes the excitement has passed off, and the Indian grows calm, but appears exhausted. Were he at home in his hut, he would sleep off the remaining fumes, but now he must shake off his drowsiness by renewing the dance. . . .

White men who have partaken of *caapi* in the proper way concur in the account of their sensation under its influence. They feel alternations of cold and heat, fear and boldness. The sight is disturbed, and visions pass rapidly before the eyes, wherein everything gorgeous and magnificent they have heard or read of seems combined; presently the scene changes to things uncouth and horrible. These are the general symptoms, and intelligent traders of the Upper Rio Negro, Uaupés, and Orinoco have all told me the same tale, merely with slight personal variations. A Brazilian friend said that when he once took a full dose of *caapi* he saw all the marvels he had read of in the *Arabian Nights* pass rapidly before his eyes as in a panorama; but the final sensations and sights were horrible, as they always are. . . .

Aya-huasca is used by the Zaparos, Angutéros, Mazánes, and other tribes precisely as I saw *caapi,* used on the Uaupés, viz. as a narcotic stimulant at their feasts. It is also drunk by the medicine-man, when called on to adjudicate in a dispute or quarrel—to give the proper answer to an embassy—to discover the plans of an enemy—to tell if strangers are coming—to ascertain if wives are unfaithful—in the case of a sick man to tell who has bewitched him, etc.

All who have partaken of it feel first vertigo; then as if they rose up into the air and were floating about. The Indians say they see beautiful lakes, woods laden with fruit, birds of brilliant plumage, etc. Soon the scene changes; they see savage beasts preparing to seize them, they can no longer hold themselves up, but fall to the ground. At this crisis the Indian wakes up from his trance, and if he were not held down in his hammock by force, he would spring to his feet, seize his arms, and attack the first person who stood in his way. Then he becomes drowsy, and finally sleeps. If he be a medicine-man who has taken it, when he has slept off the fumes he recalls all he saw in his trance, and thereupon deduces the prophecy, divination, or what not required of him. Boys are not allowed to taste *aya-huasca* before they reach puberty, nor women at any age: precisely as on the Uaupés.

From Richard Spruce (1817-1893) *Notes of a Botanist on the Amazon & Andes . . .* In Two Volumes. Macmillan & Co., Ltd. London. (1908) II, pp. 414-21, 424-25.

Yagé *(Banisteriopsis* spp.)

CONRAD VERNON MORTON

The following notes on the use and effect of the drug *yagé* are kindly supplied by Mr. [William] Klug [of Iquitos, Peru.]

'One of the most interesting plants found in the region of the upper courses of the Putumayo and Caquetá Rivers is the *yagé*. The Indians make a beverage from either the wild or cultivated *yagé*, boiling it in a large earthenware vessel an entire day until there is formed a sort of liquid, like the syrup of sugar cane. They add to the *yagé* the leaves and the young shoots of the branches of the *oco yagé* or *chagro panga . . .*, and it is the addition of this plant which produces the "bluish aureole" of their visions. These are like cinematograph views, and occur after about a half liter of the drink has been consumed in portions an eighth of a liter each at intervals of half an hour. Thereafter, the Indian falls into a profound sleep during which he is in a state of complete insensibility and anesthesia. During this period the subconscious activity acquires enormous intensity. The dreams follow each other with extraordinary precision and clearness, giving to the intoxicated person, according to the observation of missionaries, the power of double vision, and of seeing things at a distance, like certain mediums in their trances. Upon awakening, he retains clearly the hallucinations and fantastic visions which he experienced in unknown regions. Perhaps this drug has the property of developing the psychic faculties. In 1919 Dr. Zerda Bayon, specialist in chemistry of plants, gave this plant the name *Telepatina.*'

'Prof. Barriga Villalba experimented upon animals with the *yageina,* which he succeeded in isolating, with the following results: If a horse has a weak dose of a few centigrams per kilogram of its weight injected into it an extreme excitation is produced, and the animal runs in all directions. The body begins to tremble and the animal maintains its equilibrium with the greatest difficulty. With a larger dose, something like twenty centigrams per kilogram, the *yageina* becomes a real poison, and the animal loses its equilibrium, cries, falls into convulsions, its temperature is lowered, and anesthesia becomes general. The same results were obtained with dogs, in which complete anesthesia without loss of vision or sense of smell was proved.'

'The small dose which Barriga Villalba tried upon himself produced a profound sleep and certain sensations of well-being. But this was very far from being the effect on the savages, for which reason Professor Muñoz, of Colombia, employed 30 to 40 grams of the drink, prepared according to the manner of the natives. Effect: At first there was a slight dilation of the pupils. All exterior objects acquired a strange appearance, aureoled and of a blue color. Then came the most extraordinary hallucinations, resembling those of hashish, very magnificent, very terrifying. These are due without doubt to the excitation of the cerebral centers of visions, the sensibility of which is such, that the person who has taken *yagé* is capable of *seeing objects in the midst of the most complete obscurity.*'

'In Umbria, I have had occasion to converse with persons of education who have told me of taking *yagé*, prepared by the savages (but without the addition of the leaf of *chagro panga* or *oco-yagé* for the cure of malaria from which they suffered, and they have assured me that with three drinks of this (about 150 grams) they have been cured completely, and that for several years they have not suffered further from this illness.'

From C. V. Morton "Notes on Yagé, a Drug Plant of Southeastern Colombia." Academy of Sciences. *Journal.* Washington. 21 (1931) #20, pp. 487-88.

Natéma, Aya-huasca *(Banisteriopsis* spp.)

RAFAEL KARSTEN

. . . The two other principal narcotic drinks [in addition to tobacco] of the Jibaros and the Canelos [sub-tribe, Ecuador] Indians are prepared from the vine *Banisteria caapi (natéma,* Jibaros), *(aya-huasca,* Quechua). . . .

At about 8 o'clock the principal ceremony of the second day [of the "Victory Feast"], the drinking of the sacred *natéma* takes place. When the drink is prepared for the feast the slayer [of the enemy who brings the head trophy] himself has to assist in order to transfer to it the supernatural power with which he is believed to be invested.

For the preparation of the *natéma* some pieces of the stem of the vine are cut off, crushed with clubs, and parted into thinner fibres, which are boiled in water for a couple of hours. The fibres are then taken out and the drink is ready. The *natéma* drink, which has the effect of producing in the drinker peculiar visions and hallucinations which are ascribed by the Indians to certain spirits, is generally mixed with some tobacco water, by which its narcotic effects are increased. [The preparing and drinking of the *natéma* at this feast is done with certain ceremonies.]

Some pieces of the stem of the *natéma* plant are laid on a banana leaf on the ground. Upon another banana leaf a large and a small wooden club are laid, and with these the *natéma* stems are to be crushed. There is besides a pot in which the drink will be cooked. The priest as usual gives the slayer tobacco juice through the nose. Then he grasps him by the wrist and makes him seize a club with his right hand and a *natéma* stem with his left and crush the stem, laying it upon the other club. The slayer divides the stem into three or four fibres and puts them down into the pot, the priest holding his hand. Another Indian without ceremony crushes some other pieces of the *natéma* stem and arranges them in a ring within the pot. The slayer, whose hand is held by the priest, now pours some water into the pot from a water bottle and places the pot on the fire, some other men attending to the pot while it boils.

At another fire close by tobacco is simultaneously being boiled, to be mixed with the *natéma*. A small clay pot is placed by the fire, and some leaves of tobacco are laid on a banana leaf at the side of it. The slayer, whose hand is held by the priest, takes a tobacco leaf and carefully puts it, first on the edge of the pot and then into it. . . . [The same is done by the slayer's wife and daughter.] Thereupon the slayer, assisted by the priest, pours some water into the pot and places it on the fire.

On the spot where the *natéma* is cooked there is also placed a piece of the stem of the manioc plant and two narrow strips of the bark of a tree which the Jibaros call *samiki*. The slayer takes one of the strips, winds it round his index finger, and ties it into a ring of the same size as the finger. By means of one end of the strip which after the tying of the ring has been left free, he then attaches the ring to the piece of the manioc stem. Thereupon he in the same way makes another ring of the other strip of the size of his index finger, and attaches it to the manioc stem at the side of the first ring. With the aid of the priest he ultimately places the manioc stem, with the two rings attached to it, upon the tobacco pot boiling on the fire in such a way that it rests upon the edges of the pot.

The object of this ceremony is to establish a mysterious connection between the slayer, who, as we have seen, is supposed to be filled with supernatural power, the tobacco water to be mixed in the *natéma* drink, and the manioc plant which the persons drinking of the narcotic will see in the dream. The two rings formed of the rind of the *samiki* tree will in this connection serve as mediums for the transference of the power. The Jibaros ascribe magical virtues to the *samiki* tree itself, and a little of the bark is generally mixed with the *natéma* with a view to increasing the efficacy of the drink. The bast strips of the tree having been formed into rings of the same size as the slayer's finger, are believed to catch his power, and thus to transfer it to the manioc and the boiling tobacco pot. The persons partaking of the sacred drink are afterwards, in the narcotic sleep, supposed to see, among other things, the manioc fields of the slayer in a flourishing state and bearing a rich crop of fruit. . . .

The slayer, whose hands are held by the priest, seizes the small tobacco pot and pours its contents into the *natéma* pot. The drink is now ready to be consumed.

In the drinking of the *natéma* at the victory feast both men and women, even half-grown children, take part, all "who want to dream" being allowed to drink of the narcotic. Even the slayer, as well as his wife and daughter, drink *natéma.* The drinking has throughout a ceremonial character. A number of beautifully ornamented clay dishes are placed on the ground in two rows. The priest and two or three other old men fill them with *natéma* and give them to the persons who are going to drink. Before they give the dish to a man or woman they each time sing a long conjuration over it, summoning the *natéma* spirits. The person who received the dish quickly empties its contents, which amount to nearly half a litre. Immediately thereafter he or she goes out and throws up the quantity drunk (for the *natéma* at first has the effect of an emetic). Then the person again enters the house, again empties a *pininga* of *natéma,* which is given him by an old man with the same ceremony as before, and immediately again throws it up. The same process is repeated a third time. Each person who drinks *natéma* at the Tsantsa [victory]-feast thus has to empty three dishes of the sacred drink.

The persons who drink *natéma* have not previously eaten or drunk anything, and afterwards also they have to fast strictly until they have slept and dreamed. The majority of those who have drunk the narcotic leave the house and go out to sleep in some shelters of palm leaves made in the forest at a short distance from the house. Most of these are men, but at least one of them ought to be a woman, a female relative of the slayer. The slayer and his wife and daughter, who also have drunk *natéma,* do not leave the house, but have their dreams inside. The dreamers remain in the forest sleeping until the afternoon. Then they take a bath in the river and return to the house, where they tell the older Indians what kind of dreams and visions they have had. Now they are also allowed to break their fast. Their food consists only of a dish of boiled and mashed manioc and boiled ripe bananas. The dreamer has to receive the dish containing the food from the hand of the same old man who in the morning had given him or her the *natéma.*

The object of the drinking of the *natéma* at the victory-feast is to ascertain whether everything will turn out favorably for the slayer in the future, whether he will have a long life, attain material prosperity and be lucky in his undertakings. The slayer, as well as his nearest relatives who have drunk *natéma,* will see in the dream his house surrounded by large and flourishing plantations of manioc and bananas; they will see his domestic animals, his swine and his hens, numerous and fat, etc. But at the same time the persons who have drunk the sacred drink will be benefitted themselves, being purified from impure and disease-bringing matter, and gaining strength and ability for their respective work and occupations. . . .

From Rafael Karsten (1879-1956) "The Head-Hunters of Western Amazonas. The Life and Culture of the Jibaro Indians of Eastern Ecuador and Peru." Helsingfors. Societas Scientiarum Fennica. *Commentationes Humanarum Litterarum.* VII (1935) #1 pp. 124, 343-45.

Natéma *(Banisteriopsis caapi)*

MICHAEL J. HARNER

When I first undertook research among the Jívaro in 1956-57, I did not fully appreciate the psychological impact of the *Banisteriopsis* drink upon the native view of reality, but in 1961 I had occasion to drink the hallucinogen in the course of field work with another Upper Amazon Basin tribe. For several hours after drinking the brew, I found myself, although awake, in a world literally beyond my wildest dreams. I met bird-headed people, as well as dragon-like creatures who explained that they were the true gods of this world. I enlisted the services of other spirit helpers in attempting to fly through the far reaches of the galaxy. Transported into a trance where the supernatural seemed natural, I realized that anthropologists including myself, had profoundly underestimated the importance of the drug in affecting native ideology. Therefore, in 1964, I returned to the Jívaro to give particular attention to the drug's use by the Jívaro shaman.

The use of the hallucinogenic *natéma* drink among the Jívaro makes it possible for almost anyone to achieve the trance state essential for the practice of shamanism. Given the presence of the drug and the felt need to contact the 'real', or supernatural, world, it is not surprising that approximately one out of every four Jívaro men is a shaman. Any adult, male or female, who deserves to become such a practitioner, simply presents a gift to an already practicing shaman, who administers the *Banisteriopsis* drink and gives some of his own supernatural power—in the form of spirit helpers or *tsentsak*—to the apprentice. The spirit helpers, or 'darts', are the main supernatural forces believed to cause illness and death in daily life. To the non-shaman they are normally invisible, and even shamans can perceive them only under the influence of *natéma*. . . .

He had drunk, and now he softly sang. Gradually, faint lines and forms began to appear in the darkness, and the shrill music of the *tsentsak,* the spirit helpers, arose around him. The power of the drink fed them. He called, and they came. First, *pangi,* the anaconda, coiled about his head, transmuted into a crown of gold. Then *wampang,* the giant butterfly, hovered above his shoulder and sang to him with its wings. Snakes, spiders, birds and bats danced in the air above him. On his arm appeared a thousand eyes as his demon helpers emerged to search the night for enemies.

The sound of rushing water filled his ears, and listening to its roar, he knew he possessed the power of *tsungi,* the first shaman. Now he could see. Now he could find the truth. He stares at the stomach of the sick man. Slowly, it became transparent like a shallow mountain stream, and he saw within it, coiling and uncoiling, *makanchi,* the poisonous serpent, who had been sent by the enemy shaman. The real cause of the illness had been found.

From Michael J. Harner "The Sound of Rushing Water. A Hallucinogenic Drug Gives the Jívaro Shaman Entrance to the 'Real World' and gives him the power to Cure or Bewitch." Reprinted by permission from *Natural History* Magazine, June-July 1968; pp. 28-29. Copyright (c) The American Museum of Natural History, 1968.

Aya-huasca *(Banisteriopsis* spp.)

ARA H. DER MARDAROSIAN & OTHERS.

Of major importance among the non-food plants used by the Cashinahua [a small tribe of the tropical rain forest, Amazon drainage basin of eastern Peru and western Brazil] are several species of the genus *Banisteriopsis,* which, along with several species of *Psychotria* [*Rubiaceae* or Madder Family] is used to prepare a brew called *nixi pae,* "vine drunkenness", and known to the local Peruvians as *ayahuasca.* . . .

Ayahuasca is drunk by the Cashinahua in order to gain information not available through the normal channels. The hallucinations are thought to be the experiences of one's dream spirit; they are portents of the future or reminders of the past, and through them the drinkers are able to learn of things, persons, and events removed from them by time and/or space. . . .

Although all the men know how to prepare the beverage, it normally falls to one of two men from each village to make the preparations. The host goes to the jungle and without any ritual or ceremony selects and cuts one to two meters of *Banisteriopsis* and three to five branches of *Psychotria.* On returning to his house, he cuts the vine into 6- to 8-inch segments, which he pounds lightly with a rock and places in a clay cooking pot with a 2- to 4-gallon capacity. The leaves and buds of the *Psychotria* are stripped from the branches and added to the pot, which is then filled with water. A fire is lit around the base of the pot and allowed to burn until the water nearly reaches a boil. The brew is steeped for about an hour, after which it is ladled off into smaller pots to cool.

As the hour for the affair approaches, the host places some stools and logs near the hearth. Each man, when he arrives, goes to the pots and dips out about one pint of the liquid. He sings or chants several phrases over the brew asking it to show him many things and then gulps it down. He then joins the others and talks or chants quietly while waiting for the effects of the drug to begin. After fifteen minutes he may drink another pint, particularly if he wishes to hallucinate freely, or as they say "to have a good trip".

Once the drug "begins to shake them", chanting begins in earnest. Each man sings independently. Chants often involve conversations with the spirit of *ayahuasca;* at other times, they merely consist of the rhythmic repetition of the monosyllabic *'e 'ee e' e' e.* Those who do not know the chant sit next to someone who does and sway their bodies in time with the rhythm.

Although each man operates on his own, the group is very important, as it provides him a contact with the real world, without which the terrors of the spirit world through which he is travelling would be overwhelming. Frequently, a group of men will line up on a log, each one wrapping his arms and legs around the man ahead of him. Only the men who are "strong", i.e., those who have had many years of experience with *ayahuasca,* will not maintain physical contact with at least one other person.

The volume of the chanting rises and falls, punctuated by shrieks of terror, wretching, and vomiting. No attempt is made to coordinate either the rhythm or the pitch of the chants. Each man devotes his attention to what he is experiencing and his own search for knowledge.

In spite of the individual nature of the hallucinogenic experience, there is a high degree of similarity in the content and frequency of occurrence of particular hallucinations from individual to individual during any one night of drinking. Certain themes also recur every time they drink *ayahuasca.* The most frequent of these are: (1) brightly colored, large snakes, (2) jaguars and ocelots, (3) spirits, both of *ayahuasca* and others, (4) large trees, often falling trees, (5) lakes, frequently filled with anacondas and alligators, (6) Cashinahua villages and those of other Indians, (7) traders and their goods, and (8) gardens. . . .

All informants speak of a sense of continual motion and rapid change, or, as they say, transformation. Particular hallucinations wax and wane, interspersed by others in a very fluid manner. There is a sense of darkness interrupted often by flashing bright colors or brightness when the horizon seems to collapse. Time and space perceptions are distorted.

The following is an excerpt from a longer tape-recorded report by one of the men who drank *ayahuasca* on August 27, 1968. . . .

"We drank *nixi pae.* Before starting to chant, we talked a bit. The brew began to move me and I drank some more. Soon I began to shake all over. The earth shook. The wind blew and the trees swayed. . . . The *nixi pae* people began to appear. They had bows and arrows and wanted to shoot me. I was afraid but they told me their arrows would not kill me, only make me more drunk. . . . Snakes, large brightly colored snakes were crawling on the ground. They began to crawl all over me. One large female snake tried to swallow me, but since I was chanting she couldn't succeed. . . . I heard armadillo tail trumpets and then many frogs and toads singing. The world was transformed. Everything became bright. I moved very fast.

Not my body but my eye spirit . . . I saw lots of gardens full of manioc and plantains. The storage sheds were full of corn. The peanut racks were full. . . . I came down the trail to the village. There was much noise, the sound of people laughing. They were dancing *kacha,* the fertility dance. Everybody was laughing. Many of the women were pregnant. I was happy. I knew we would be well and have plenty to eat."

From Ara H. Der Marderosian, Kenneth M. Kensinger, Jew-Ming Chao and Frederick J. Goldstein. "The Use and Hallucinatory Principles of a Psychoactive Beverage of the Cashinahua Tribe (Amazon Basin)." National Institute of Mental Health. 5454 Wisconsin Avenue, Chevy Chase, Md. 20015 National Clearing House For Mental Health Information. *Drug Dependence.* (October, 1970) Issue #5, pp. 7-8. (Reprinted by permission)

Yajé *(Banisteriopsis caapi)*

GERARDO REICHEL-DOLMATOFF

The drinking of *yajé* represents a return to the maternal womb, to the source and origin of all things. The partaker 'sees' all the tribal divinities; the creation of the universe, of the first human beings, and of the animals; and the establishment of the social order, especially regarding the law of exogamy. The Indians claim to see not only abstract designs but also the figures of people and animals, such as jaguars, alligators, snakes, and turtles, in complex mythological scenes. Having had this experience, the individual is firmly convinced of the verity of his religious beliefs. But the return to the womb is also an acceleration of time, and is equivalent to death. Upon drinking the *yajé*, the individual 'dies', but later revives in a state of great wisdom. At the same time, the hallucinatory experience is symbolically a sexual act, essentially incestuous, in which he returns momentarily to the mythical stage of Creation.

From G. Reichel-Dolmatoff "Notes on the Cultural Extent of the Use of Yajé *(Banisteriopsis caapi)* among the Indians of the Vaupés, Colombia." *Economic Botany.* Lancaster, Pa. 24 (1970) p. 33. (Reprinted by permission)

Myristicaceae or Nutmeg Family

Hakúdufha *(Virola* spp.)

THEODOR KOCH-GRUENBERG

[Conjuring and Healing among the Yekwana Indians of the Head Waters of the River Orinoco.]

Of an especial magical importance are cures, during which the witch-doctor inhales *hakúdufha.* This is a magical snuff used exclusively by witch-doctors and prepared from the bark of a certain tree which, pounded up, is boiled in a small earthware pot, until all the water has evaporated, and a sediment remains at the bottom of the pot. This sediment is toasted in the pot over a slight fire and is then finely powdered with the blade of a knife. Then the sorcerer blows a little of the powder through a reed . . . into the air. Next, he snuffs, whilst, with the same reed, he absorbs the powder into each nostril successively. The *hakúdufha* obviously has a strongly stimulating effect, for immediately the witch-doctor begins singing and yelling wildly, all the while pitching the upper part of his body backwards and forwards. . . .

From Theodor Koch-Gruenberg (1872-1924) *Vom Roroima zum Orinoco. Ergebnisse einer Reise in Nordbrasilien und Venezuela in den Jahren 1911-1913.* (Translations: Results derived from travelling in North Brazil and Venezuela during the years 1911-13.) 5 Volumes. Strecker & Schroeder Stuttgart. (1917-28) III, p. 386.

Epéna *(Virola* spp.)

GEORG J. SEITZ

We talked with a young Indian of the Kavaretari tribe who had learned Portuguese in the mission school at Tapuruquara.

We gathered some explanations about the ceremony from his conversation:

We asked 'Do you snuff *epéna?*' He answered, 'No, I am not allowed to. I am not grown up yet!'

'When will you grow up?', we then asked. 'I don't know', he said, 'but I think it will be soon.'

Next question: 'Who decides when you are grown up?' 'My father. He shows me how to make *epéna* powder, and tells me what happens when I sniff it.'

Question: 'What will happen then?'

Answer: 'Then I will see the *häkula,* who are big men living there above in big huts.'—He pointed to the sky and continued: 'The *epéna* makes me so big that I can see them and talk with them!' . . .

Another Indian named Daniel, who had lived in the Tapuruquara-mission for some years before returning to the tribe and marrying, told me that he had seen 'angels' while under the effect of *epéna.* And that he had talked with them!

The Indian feels that he is a giant; everything around him takes enormous and magnificent forms. In the midst of a super-dimensional world, he feels, like a superman! Consequently, his movements correspond to his state of excitation. These are braggart's gestures. These symptoms are accompanied by profuse salivation, a bad headache, a fixed stare and heavy perspiration. The symptoms reveal a state of strong intoxication.

The second effect is imagined. The Indian sees things that he has been taught to see. One sees 'big men' because his father has told him that he would. The other saw 'angels' because he had been taught in the mission that they are more powerful than *häkula*—spirits!

From Georg J. Seitz "Epéna, the Intoxicating Snuff Powder of the Waika Indians and the Tukano Medicine Man, Agostino." In: (D. H. Efron, ed.) *Ethnopharmacologic Search for Psychoactive Drugs.* U. S. Government Printing Office. Washington, D.C. (1967) (Public Health Service Publication #1645). p. 334. (Reprinted by permission)

Nyakwana (*Virola* spp.)

RICHARD EVANS SCHULTES

. . . The effects of *Virola* intoxication vary, but amongst the Indians, they usually include initial excitability—setting in within several minutes from the first snuffing—numbness of the limbs, twitching of the facial muscles, inability to co-ordinate muscular activity, nausea, visual hallucinations and finally, a deep, disturbed sleep. Macroscopia is frequent and enters into Waika [inhabiting the very headwaters of the Orinoco in Venezuela and the Brazilian territory north of the Rio Negro] beliefs about the spirits that dwell in the plant. A description of my own intoxication indicates several points of interest. 'The dose was snuffed at five o'clock. Within fifteen minutes a drawing sensation in the forehead gave way to a strong and constant headache. Within a half hour, the feet and hands were numb and sensitivity of the fingertips had disappeared: Walking was possible with difficulty, as with beri-beri. I felt nauseated until eight o'clock and experienced lassitude and uneasiness. Shortly after eight, I lay down in my hammock, overcome with drowsiness, which, however, seemed to be accompanied by a muscular excitation except in the hands and feet. At about nine thirty, I fell into a fitful sleep which continued, with frequent awakenings, until morning. The strong headache lasted until noon. A profuse sweating and what was probably a slight fever persisted throughout the night. The pupils were strongly dilated during the first few hours of the intoxication. No visual hallucinations nor colour sensations were experienced.'

From Richard Evans Schultes "The Plant Kingdom and Hallucinogens (Part II)" *Bulletin on Narcotics*. (United Nations. Division of Narcotic Drugs. Department of Economic and Social Affairs). New York. 22 (1970) #1, pp. 19-22. (Reprinted by permission)

Solanaceae or Nightshade Family

Tlápatl & Tzitzintlápatl *(Datura* spp.)

BERNARDINO DE SAHAGUN

On Certain Intoxicating Herbs.

. . . There are other shrublike plants, called *tlápatl.* These plants grow spineless heads of lemon-shape. They have a green bark, slightly broad leaves, white flowers, and black, fetid seeds; they have a disgusting taste, and if eaten, will intoxicate and cause permanent madness. The seeds are useful against gout if placed in form of an ointment on painful spots. The odor of these plants is as harmful as are their seeds.

There are other plants of this kind. The Indians call these *tzintzintlápatl,* because they have spiny heads and the same properties as the foregoing plants.

From Fray Bernardino de Sahagún (1495-1590) *Historia General de las Cosas de Nueva España.* (Translation: General History of Things of New Spain (Mexico). 5 volumes. Pedro Robredo. México. D.F., (1938) Book XI, ch. vii, paragraph 1.

Tlápatl *(Datura* sp.)

FRANCISCO HERNANDEZ

Of the *Tlápatl,* Stramonium.

The *tlápatl* grows white and ramified roots; its fruit is round and resembling a sea-urchin; its stem is green; its leaves are like those of the grape-vine, broad, but more deeply lobed; its flower is white, long and shaped like a capsule. The *tlápatl* grows everywhere, but principally in Tepecuacuilla and México. It is of cold temperament and without any noticeable taste or odor. A decoction of the leaves used as ointment on the body is against fever, particularly against ague; it also is applied in form of a suppository or of a small ball. The fruit and leaves are good against pains in the chest. Mixed with water and instilled in the ears, they alleviate deafness. Placed on the pillow of sleepless persons, it induces sleep; but if it were taken somewhat abundantly, it would cause madness.

From Dr. Francisco Hernandez (1517/18-1587) *Nova Plantarum, Animalium et Mineralium Mexicanum Historia* . . . (Added title page: *Rerun Medicarum Novae-Hispaniae Thesaurus* . . .) (Translation: New History of Mexican Plants, Animals and Minerals. Thesaurus of Medical Matters of New Spain (Mexico) . . . Typis Vitalis Mascardi, Romae. (1651) Book VIII, ch. xxiix (i.e., xxviii).

Chamico *(Datura* spp.)

BERNABÉ COBO

The plant called *chamico* by the Indians, grows to a height of about one cubit. The leaves, bearing some notches at their edges, resemble those of the wild amaranth. Its blossoms, resembling bell flowers, become small heads the size of a poppy fruit. These capsules, covered with sharp thorns, enclose the seeds. The seeds resemble those of the radish, and the Indians say that they are as cold as celery. If a person drinks a decoction of these seeds, his senses will be benumbed. The Indians use this drink for intoxication, but he who takes too much of this brew will so much be deprived of his reason that, though he has his eyes open, he cannot see or comprehend anything. The Indians formerly wrought great harm with the brew. Not long ago, it happened in the Kingdom [of Peru] that a man whom I knew was walking with a companion who, having plans to rob my friend, gave him a *chamico* drink. This caused the man to lose his mind and to become so enraged that, naked for his shirt, he started to throw himself into the river. He was seized like a mad man and restrained. For two days, he remained in this condition without regaining consciousness. The juice of the *chamico* leaves mixed with a few drops of vinegar and applied over the liver and spine, cures hot, distempered conditions; it is also effective against high fevers; and, if taken regularly, a decoction of the same leaves, relieves a steady fever.

From Padre Bernabé Cobo (1582-1657) *Historia del Nuevo Mundo.* (Written in 1653). Four volumes. E. Rasco, Sevilla. (1890-93) Book IV, ch. c. (First edition).

Borrachero *(Datura candida)*

JUAN DE CASTELLANOS

[The chiefs of the Chibcho Indians on the middle Magdalena River, personally prepare the tombs for their lords at most concealed places, the location of which they keep very secret.] Some of these graves are in the woods and thickets, others on high mountains. Sometimes some of these tombs are covered with water taken from rivers or lakes and these are the most hidden ones. . . . They make these graves very deep and on the lowest level they place the [bodies of] the kings, seated on the so-called *duhos* made mostly of gold and richly ornamented. [The king is seated in the tomb adorned with his jewels, fully armed and provided with food and drink.] This . . . death chamber they cover with a layer of earth on which they place about three or four unfortunate wives of the king, his most loved ones. These women are buried alive, and on top of them [on another layer of earth, are buried,] also alive, the master's slaves, those who had served him best. Finally, a last layer of earth closes this lugubrious and hateful sepulcher. . . . Before being brought to this monstrous cave, that they will not suffer any fright or pangs of death, the women and slaves are given by the chiefs a certain potion mixed into their ordinary beverage [*chicha,* corn wine]. This potion is prepared from intoxicating *tabaco* and from other leaves of a tree which they call *borrachero.* And thus their senses become so benumbed that they will not feel any pain.

From Juan de Castellanos (1522-1607) *Historia del Nuevo Reino de Granada.* (Translation: History of the New Kingdom of Granada (i.e. Colombia). 2 volumes. A. Pérez Dubrull. Madrid. (1886-87) I, pp. 64-66. (First edition)

Figure 8.
One of the atrophied "races" of *Datura candida* which the Kamsa Indians reproduce vegetatively for use as a source of a narcotic drink. Sibundoy, Colombia. Photograph courtesy of R. E. Schultes.

Floripondium *(Datura candida)*

JOSEF SKINNER

. . . These Indians likewise admit an evil being, the inhabitant of the centre of the earth, whom they consider as the author of their misfortunes, and at the mention of whose name they tremble. The most shrewd among them take advantage of this belief, to obtain respect; and represent themselves as his delegates. Under the denomination of *Mohanes,* or *Agoreros,* they are consulted even of the most trivial occasions. They preside over the intrigues of love, the health of the community, and the taking of the field. Whatever repeatedly occurs to defeat their prognostics, falls on themselves; and they are wont to pay their deceptions very dearly. . . .

But the principal power, efficacy, and, it may be said, misfortune, of the *Mohanes,* consist in the cure of the sick. Every malady is ascribed to their enchantments, and means are instantly taken to ascertain by whom the mischief may have been wrought. For this purpose, the nearest relative takes a quantity of the juice of *floripondium* [*D. Candida*] and suddenly falls, intoxicated by the violence of the plant.
he is placed in a fit posture to prevent suffocation, and on his coming to himself, at the end of three days, the *Moharis* who has the greatest resemblance to the sorcerer he saw in his visions, is to undertake the cure, or if, in the interim, the sick man has perished, it is customary to subject him to the same fate. When not any sorcerer occurs in the visions, the first *Mahoris* they encounter has the misfortune to represent his image. . . .

From [Joseph Skinner] *The Present State of Peru.* Richard Phillips. London. (1805) pp. 275-76.

Maíkoa *(Datura candida)*

RAFAEL KARSTEN

Equal in importance to the *natéma (aya-huasca)* and still strong in its effects is the narcotic which is called *maîkoa* by the Jibares and *huántuc* by the Canelos Indians. It is prepared from a shrub . . . [*Datura candida*] of the family *Solanaceae*. Its poisonous properties may be compared with those of belladonna or opium, and even seem to surpass them. Insensibility, hallucinations, and even temporary madness seem to be the regular effects produced by this narcotic drug. . . .

In order to prepare the *maîkoa*-drink, the Jibaro Indian scratches off some of the light green rind of the stem, and squeezes out the juice with the hand into a gourd until he gets the desired quantity, about a small drinking-glass full (about 200 gr.). This is the dose which is taken at one time by one person and seems to be sufficient to produce the effects desired.

Thus, *maîkoa* is not prepared by cooking like *natéma*, but simply consists of the juice contained in the rind. It is of a light green colour and not bitter to the taste like *natéma*, but rather insipid. Its effects, as already mentioned, are much stronger and show sooner than those of the other narcotic. Young people, when they take *maîkoa* for the first time, often become mad and strike out with weapons, sticks, and whatever they happen to get hold of, and may then become dangerous to those around. Therefore, when a Jibaro youth drinks *maîkoa*, some older men are always present who watch his movements and sometimes even tie him down. The drugged man begins to talk confusedly and incoherently about all sorts of visions which he imagines he has, about spirits whom he fancies he meets, and at last falls into a deep, lethargic sleep, "just as if he were dead" *(hakáma)*.

Maîkoa is drunk by the Jibaros on much the same occasions as the *natéma*, and the revelations which they imagine they receive are much the same; yet to a still higher degree than the latter narcotic, *maîkoa* is the particular drink of warriors. If one asks a Jibaro Indian why he drinks *maîkoa,* his answer is usually: "It is in order to see whether I shall kill my enemies" . . . Similarly in other important matters, although they have no particular reference to war, the Jibaro Indian consults the spirits by drinking *maîkoa.*

The drinking ceremony takes place in the house, but the Indian prepares for it by taking tobacco water in repeated doses, and for this purpose he retires to the forest. In a remote place far from the habitations of men—sometimes the Indian has to walk one or two days to reach it—a shelter of palm leaves is constructed; it is the "dreaming hut", called *ayámdai.* It is much larger and made with greater care than the ordinary Indian shelter, and outside it there is an open place where the ground is carefully cleared and made level. On this open place the spirits appear *dancing* in all sorts of terrible shapes, and first of all come those powerful demons whom the Jibaros call *arútama,* "the Old Ones", who give the sleeping Indian information and advice in the matters which interest him. . . .

The Jibaros are in the habit of retiring to the forest on these occasions because the *arútama* are supposed to appear to them more easily there. Generally the Indian who wants to dream remains away for three or four days. He may go alone, or he may go in company with other men. While on the way he has to fast strictly, but instead of food he takes tobacco water in repeated doses, through the nose. As soon as he reaches the shelter, he first takes a bath, standing in the pool [of a waterfall nearby] and allowing the streaming water to fall on his naked shoulders. The bath will prepare him favourably for the sleep and the revelations of the spirits. These spirits, it should be understood, are demons of the forest, but they are also present in the water of the cascades. After the bath the Indian goes to the hut, where in the evening he again prepares tobacco water, boiling the leaves in a small earthen pot which he has brought with him for the purpose. Having taken some of the medicine he falls in a deep sleep.

In the following morning he again takes a bath in the pool. During the whole day he stays in the shelter, fasting strictly, a roasted plantain being his only food. On the other hand, he goes on taking tobacco water. Living this way he spends a couple of days in the forest. On the fourth day, having taken his last bath in the waterfall, he returns home where he tells his dreams to the other Indians. It is only then that he takes *maîkoa,* of which one dose alone, after his nervous sensibility has been enhanced by the tobacco-drinking and the fasting, will infallibly produce in him the usual strange physical and psychical effects.

Since among the Jibaros the education of the boys chiefly aims at making them good warriors, *maîkoa*-drinking also occurs as a puberty ceremony. . . .

. . . *maîkoa* . . . may also be consumed by women, . . . On these occasions the ancestral spirits, the *arútama,* are supposed to appear to them, and especially the Earth-mother *Nungüi.* When appearing to women these spirits . . . give them instruction and information . . . about domestic work and occupations which concern the women, in the same way as when they drink *natéma.*

From Rafael Karsten (1879-1956) "The Head-Hunter of Western Amazonas. The Life and Culture of the Jibaro Indians of Eastern Ecuador and Peru." Helsingfors. Societas Scientiarum Fennica. *Commentationes Humanarum Litterarum* VII (1935) #1, pp. 438-41.

Nacazcul or Toloatzin [*Toloaché*]

FRANCISCO HERNANDEZ

(Datura inoxia=Datura meteloides)
Of the *Nacazcul,* also called *Toloatzin* [Toloaché]

The *nacazcul,* by some called *toloatzin,* is a kind of *tlápatl.* It grows in the province of *Huexotcinco.* Its fruit first is spiny, but later loses its spines; it is round and divided in four parts like a melon. Its seed is tawny and similar to that of the radish. The *nacazcul* grows everywhere on waste places or on hedges in Pahuatlan. Its seed, when dried, ground and mixed with resin, excellently solders and sets broken bones or restores dislocated. For this purpose the Indians place on them birds' feathers as splints and then take the patient to a hot vapor bath, called *temazcalli* in their native language. This cure is repeated as often as they deem it necessary. Four crushed leaves taken in water are used against pains in the whole body, even against those deriving from the French disease. Mixed with yellow capsicum it is used as an ointment by the Indians, but care must be taken that not an excessive amount of it were used, for this would produce an alienation of the mind, visions and deliriums.

From Dr. Francisco Hernandez (1517/18-1587) *Nova Plantarum Animalium et Mineralium Mexicanorum Historia . . .* (Added title page: *Rerum Medicarum Novae-Hispaniae Thesaurus . . .)* Typis Vitalis Mascardi, Romae. (1651) Book IV, ch. xviii.

U'teaw ko'hanna (The white flowers) *(Datura inoxia)*

MATILDA COXE STEVENSON

There can be no question as to the early use of antiseptics and narcotics by the Zuñi . . . The Zuñi rain priests administer *Datura meteloides,* [i.e. *D. inoxia*] that one may become a seer, and the Zuñi "doctor" gives the root of the plant to render his patient unconscious while he performs simple operations—setting fractured limbs, treating dislocations, making incisions for removing pus, eradicating diseases of the uterus, and the like. The narcotic is seldom employed by the Zuñi for the extraction of bullets, as men, they say, are not like women, and they must be *men*. . . .

The roots and flowers of . . . [*Datura inoxia*] ground together into meal, are applied to wounds of every description . . .

Datura stramonium acts very powerfully upon the cerebro-spinal system, causing a line of symptoms showing it to be a narcotic-irritant of high degree. The symptoms collated from many cases of poisoning by this drug are: vertigo, with staggering gait, and finally unconsciousness; stupor and deep sleep, with stertorous breathing; mania, with loquaciousness or melancholia; hallucinations of terrifying aspect, the patient bites, strikes and screams, and throws the arms about, or picks and grasps at unattainable objects; congestive headaches, with dull beating and throbbing in the vertex. The pupils are dilated, and the patient suffers from photophobia, diplopia, and hemeralopia; the eyes are wide open, staring and set, or are contorted, rolling, and squinting. The face becomes red, bloated, and hot, the mouth spasmodically closed, and the tongue dry and swollen; the patient suffers greatly from thirst, but the sight of water throws him into a spasm and causes great constriction of the throat, foaming at the mouth, but seldom vomitting. The sexual functions are often excited, more especially in women, in whom it causes nymphomania. Spasms of the muscles of the chest are of frequent occurrence; inspiration is slow and expiration quick. Paralysis of the lower limbs and loss of speech, with twitchings and jerkings of the muscles often mark a case. Its action will be seen to be similar to that of Belladonna, yet differing in many respects.

A minute quantity of the powdered root [of *Datura inoxia*] is put into the eyes, ears, and mouth of each of the *A'shiwanni* [rain priests] when they go at night to ask the birds to sing for rain. "The birds are never afraid to tell the *A'shiwanni* that they will sing when they have the powder in their eyes, ears, and mouths.". . .

A small quantity of the powdered root of *Datura* [*inoxia*] is administered by a rain priest to put one in condition to sleep and see ghosts. This procedure is for rain, and "rain will surely come the day following the taking of the medicine, unless the man to whom it is given has a bad heart."

Frequently when a man has been robbed and wishes to discover the thief, he summons to his aid a rain priest, who prepares plum offerings [of certain birds], . . . and plants them at sunrise of the day he is to treat the man who has lost his property. . . .

When the rain priest arrives at the home of the man whom he is to treat, he finds him seated in darkness in an inner room. He wears a white cotton shirt and trousers. His hair is dressed in the usual style. He has new blue woolen leggings, but he wears no shoes, nor does he have the usual headkerchief. There must be no fire in the room at this time. The rain priest, sits by the man's side, and, taking a bit of the root of . . . [*D. inoxia*] from the palm of his left hand, places it in the man's mouth, telling him to chew the medicine that he may be possessed by the power to sleep soundly and see the one who has robbed him. Then the man lies upon a pallet without speaking a word, and the rain priest retires to an adjoining room and sits by the communicating door, which he closes. He listens attentively. He must not smoke. "Should the rain priest smoke, the man could not see the robber, as A'nelakaya [a mythic boy who by the Divine Ones was transformed with his sister into the *u'teaw ko'hanna* plant, because they both knew too much of this plant's magic properties] does not like smoking at this time.

After a time the man leaves his bed and walks about the room. When he speaks, the rain priest is eager to catch every word. The man walks and lies down alternatively during the night. At daybreak the rain priest goes into the man's room and takes his arm; he may be either lying down or walking at this time. He is led into the adjoining room, and the two take seats side by side, facing eastward. The rain priest repeats to him what he has heard him say during the night, and gives him the name of the person he mentioned. The man declares that he has no recollection of what passed. After, directing him to go to the house of the one whose name he called during the night, the rain priest makes a fire, heats water, and gives the man about a quart to drink, which induces vomiting. The drinking of the water is repeated four times, each time resulting in copious vomiting, and after the last draft the root of the *Datura* is supposed to be entirely ejected. (Should) the warm water not be drunk and the medicine thereby thoroughly ejected, the flowers of *Datura* would appear over the body.) *) The man remains in his room while the rain priest goes to his own home and notifies his wife and other women of his family that a bowl containing yucca root must be carried to the house of the person whom he has treated, where yucca suds must be made and the man's head washed. During the hair washing he kneels on a blanket, and the rain priest sits back of him with a hand on each shoulder. His family may be present at this time, but they take no part in the performance. The rain priest presents four ears of corn tied together, to be planted apart from the other corn during the coming year, and the man gives a few yards of calico, or sometimes a shirt and trousers, to the rain priest, whose family bring food from his home where it has been cooking during the night, and prepare a meal. After the repast the man visits the person whom he had seen while under the influence of the *Datura* and tells him that he saw him in his dreams and knows that he stole his property. It is said that "the accused always returns the property, for he is ashamed of having been discovered."

*) The Zuni say: "When one touches a Datura blossom with moist hands, the impression will be imprinted on the hand and wherever the hand touches the body. The blossoms will appear on the hair if the hand is placed on the head."

From Matilda Coxe Stevenson (d. 1915) "Ethnobotany of the Zuñi Indians [of the extreme western part of New Mexico]." *Thirtieth Annual Report* of the Bureau of American Ethnology to the Sécretary of the Smithsonian Institution. 1908-1909. Government Printing Office. Washington. (1915) pp. 41, 47, 90-91.

Huacacachu, Yerba de Huaca or Bovachero *(Datura sanguinea,* RUIZ & PAV.)

JOHANN JAKOB VON TSCHUDI

. . . Beyond Matucanas the valley contracts into a narrow ravine no broader than the bed of the river, and it gradually assumes a wilder character. The way is difficult along the ridge of the hills which borders the left bank of the river. The vegetation is less monotonous and scanty than in the valleys of the coast, and all the fissures of the hills are filled with verdure. . . . and on the less steep declivities is seen the red thorn-apple

Figure 9.
Datura sanguinea in flower and fruit. Bogotá, Colombia.
Photograph courtesy of R. E. Schultes.

. . . To [this] the natives give the name *huacacachu, yerba de huaca,* or *bovachero;* and they prepare from its fruit a very powerful narcotic drink, called *tonga.* The Indians believe that by drinking the tonga they are brought into communication with the spirits of their forefathers. I once had an opportunity of observing an Indian under the influence of this drink. Shortly after having swallowed the beverage he fell into a heavy stupor: he sat with his eyes vacantly fixed on the ground, his mouth convulsively closed, and his nostrils dilated. In the course of about a quarter of an hour his eyes began to roll, foam issued from his half-opened lips, and his whole body was agitated by frightful convulsions. These violent symptoms having subsided, a profound sleep of several hours succeeded. In the evening I again saw this Indian. He was relating to a circle of attentive listeners the particulars of his vision, during which he alleged he had held communication with the spirits of his forefathers. He appeared very weak and exhausted.

In former times the Indian sorcerers, when they pretended to transport into the presence of their deities, drank the juice of the thorn-apple, in order to work themselves into a state of ecstasy. Though the establishment of Christianity has weaned the Indians from their idolatry, yet it has not banished their old superstitions. They still believe that they can hold communications with the spirits of their ancestors, and that they can obtain from them a clue to the treasures concealed in the *huacas,* or graves; hence the Indian name of the thorn-apple —*huacacuchu,* or grave plant.

From Johann Jakob von Tschudi (1818-1889) *Travels in Peru, during the Years 1838-1842.* Wiley & Putnam. New York, (1847) ch. x. (The original German edition was published in 1846.)

James-Town Weed *(Datura Stramonium)*

ROBERT BEVERLEY

The *James Town Weed* (which resembles the Thorny Apple of *Peru,* and I take to be the plant so call'd) is supposed to be one of the greatest Coolers in our World. This being an early Plant, was gather'd very young for a boil'd Salad, by some of the Soldiers sent thither, to pacifie the Troubles of *Bacon;* and some of them eat plentifully of it, the Effect of which was a very pleasant Comedy; for they turn'd natural Fools upon it for several Days; One would blow up a Feather in the air; another would dart Straws at it with much Fury; and another stark naked was sitting up in a Corner, like a Monkey, grinning and making Mows at them; a Fourth would fondly kiss, and paw his Companions, and snear in their Faces, with a Countenance more antick, than any in a *Dutch* Droll. In this frantick Condition they were confined, lest they should in their Folly destroy themselves; though it was observed, that all their Actions were full of Innocence and good Nature. Indeed, they were not very cleanly; for they would have wallow'd in their own Excrements, if they had not been prevented. A Thousand such simple Tricks they play'd, and after Eleven Days, return'd to themselves again, not remembering any thing that had pass'd. . . .

Wysoccan

The Solemnity of *Huskanawig* is commonly practis'd once every fourteen or sixteen years, or oftener, as their young men happen to grow up. It is an Institution or discipline which all young men must pass, before they can be admitted to be of the number of the Great men, . . . The whole Ceremony is performed after the following manner.

The choicest and brisket young men of the Town, and such only as have acquired some Treasure by their Travels and Hunting, are chosen out by the Rulers to be *Huskanawed;* and whoever refuses to undergo this Process, dare not remain among them. Several of these odd preparatory Fopperies are premised in the beginning, . . .: but the principal part of the business is to carry them into the Woods, and there keep them under confinement, and destitute of all Society, for several months; giving them no other sustenance, but the Infusion, or Decoction of some Poisonous Intoxicating Roots; by virtue of which Physick, and by the severity of the discipline, which they undergo, they become stark staring Mad: In which raving condition they are kept eighteen or twenty days. During these extremities, they are shut up, night and day, in a strong Inclosure made on purpose; one of which I saw, belonging to the *Paumaukie Indians,* in the year 1694. It was in shape like a Sugar-loaf, and every way open like a Lattice, for the Air to pass through, . . . In this Cage thirteen young Men had been *Huskanaw'd,* and had not been a month set at liberty, when I saw it. Upon this occasion it is pretended, that these poor Creatures drink so much of that Water of *Lethe,* that they perfectly lose the remembrance of all former things, even of their Parents, their Treasure, and their Language. When the Doctors find that they have drank sufficiently of the *Wysoccan,* (so they call this mad Potion) they gradually restore them to their Senses again, by lessening the Intoxication of their Diet; but before they are perfectly well, they bring them back into their Towns, while they are still wild and crazy, through the Violence of the medicine. After this they are very fearful of discovering any thing of their former remembrance; for if such a thing should happen to any of them, they must immediately be *Huskanaw'd* again; and the second time the usage is so severe, that seldom any one escapes with Life. Thus they must pretend to have forgot the very use of their Tongues, so as not to be able to speak, nor understand any thing that is spoken, till they learn it again. Now whether this be real or counterfeit, I don't know; but certain it is, that they will not for some time take notice of any body, nor any thing, with which they were before acquainted, being still under the guard of their Keepers, who constantly wait upon them every where, till they have learnt all things perfectly over again. Thus they unlive their former lives, and commence Men, by forgetting that they ever have been Boys. . . .

But the *Indians* . . . pretend that this violent method of taking away the Memory, is to release the Youth from all their Childish impressions, and from that strong Partiality to persons and things, which is contracted before Reason comes to take place. They hope by this proceeding, to root out all the prepossessions and unreasonable prejudices which are fixt in the minds of Children. So that, when the Young men come to themselves again, their Reason may act freely, without being byass'd by the Cheats of Custom and Education. Thus also they become discharg'd from the remembrance of any tyes by Blood, and are establisht in a state of equality and perfect freedom, to order their actions, and dispose of their persons, as they think fit, without any other Controul, than that of the Law of Nature. By this means also they become qualify'd, when they have any Publick Office, equally and impartially to administer Justice, without having respect either to Friend or Relation.

From Robert Beverley (c.1673-c.1722) *The History and Present State of Virginia.* Printed for R. Parker. London. (1705) Book II, paragr. 18; Book III, paragr. 31 & 33.

Latua *(Latua pubiflora)*

RODOLPHO AMANDUS PHILIPPI

Six years ago, while travelling in the Province of Valdivia, I heard that the native Indians possess a secret for maddening a person by means of a vegetable poison, and that this insanity may last longer or shorter depending on the strength of the poisonous property inherent in the plant concerned. This is a well guarded secret by the Indians. Father Romualdo, a missionary at Daglipulli, was able to ascertain that the plant is a tall shrub called *Latua* which grows here and there in the virginal forest on the coastal mountains. Finally, he even succeeded in obtaining a branch of this plant, though without leaves, for the Indians who gave it to him evidently throught that the Father was only interested in investigating the poisonous properties of the plant, principally located in the bark of the shrub. Later on I got further particulars by Mr. Juan Renous. . . . Mr. Renous could not tell me anything about the fruit of this plant, but informed me about several cases of intentional and unintentional poisoning. This poisoning may occur the more easily, as the shrub . . . greatly resembles the *Tayu,* the bark of which, externally or internally used in form of a decoction, is considered an excellent remedy for contusions, shocks caused by accidents, or horses' kicks, and so forth. Among other things

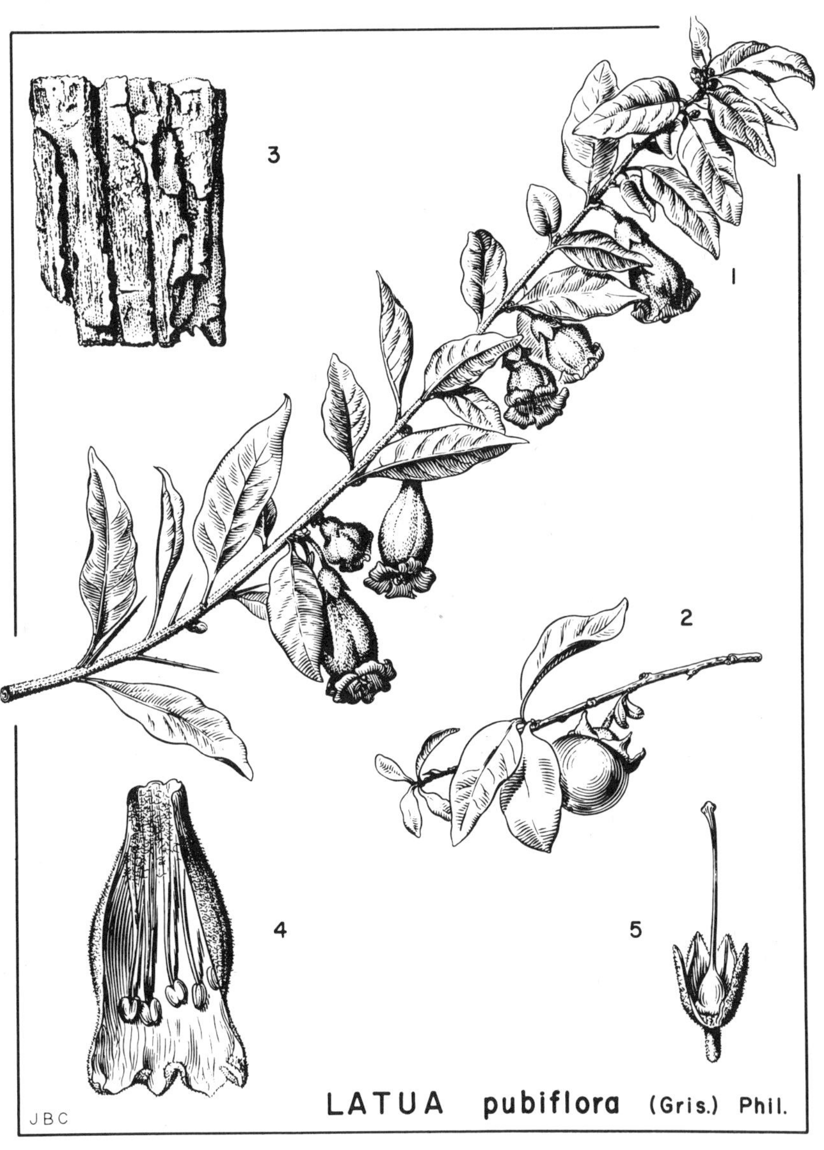

Figure 10.
Latua pubiflora. Illustration courtesy of R. E. Schultes.

Renous told me the following case: a short time ago, one of his woodcutters went to the forest to get some *Tayu* bark as remedy for a heavy blow he had received with the dull end of an axe. But instead of the *Tayu* he got hold of the *Latua* and drunk a decoction of this poison. Almost immediately afterwards the woodcutter became insane and run into the woods, where, three days later, he was found in unconscious condition. Although he recuperated within a few days, for many months he had been suffering violent headaches. . . .

From Rodolpho Amandus Philippi (1808-1904) "*Latua Ph.*, ein neues Genus der Solanaceen" *Botanische Zeitung*. Berlin-Leipzig. (Ag. 13, 1858) VI/33, pp. 241-42.

TIMOTHY PLOWMAN & OTHERS

One of the rarest and most interesting genera of the Solanaceae with . . . striking narcotic and toxic properties is *Latua,* an endemic from southern Chile with a single species: *L. pubiflora.* Owing to its great beauty and toxicity, *Latua* evoked some interest among botanists and pharmacologists shortly after its discovery towards the middle of the 19th Century— . . . Yet, *Latua* still remains relatively unknown for two reasons. First: the plant grows only in the narrow coastal cordillera between Valdivia and Chiloé, a difficult mountainous terrain with an extremely wet climate and few roads; during the rainy season, the existing roads are nearly impassable. Second: the occurrence of *Latua* and its use is a closely guarded secret surrounded by much superstition, since the plant is employed primarily by local shamans and sorcerers in their magical healing rites. Those familiar with *Latua* and its properties are very protective of this knowledge and are unwilling to discuss it with outsiders. . . .

The generic epithet *Latua* was taken from the native Mapuche name for the plant. . . .

The Spanish names of *Latua* also reveal a knowledge of its toxic properties: *palo mato,* literally "the tree that kills", meaning the same as *latué . . .; palo de los brujos . . . , arbol de los brujos . . .* and *palo de bruja,* all meaning "witches' tree".

A rare account of the action of *Latua* is given by Dr. Benkt Sparre, Curator at the Museum of Natural History of Stockholm. At the time of his self-experiment, he was Professor at the Universidad de Concepción (Chile). In a letter to the authors, he describes his experiment in the following way:

"Dec. 12, 1953—Latua pubiflora was collected at La Posada, about 3-4 kilometers north-northwest of Maullín (Llanquihue, Chile) . . ."

"According to explanations by elderly villagers of La Posada, who had not tried latue themselves, an infusion was prepared in the evening with green leaves and bark. It was said that only 'los hechiceros' (witches) used latue. Intoxicated and with an appropriate refill from a 'sub-hechicero' (witch's apprentice), they could dance and preach for a week. None of my informants had seen this, but they had heard it from old people. According to the same informants, 'los hechiceros' could quickly recover with a drug from a *Solanum* species of the section *Morella* (to which *Solanum nigrum)* belongs. Some of these species were collected nearby, but my friends could not tell me which they were. They only knew the vernacular name, 'hierba mora', which is *Solanum,* but as we know, the vernacular names mean very little. It might have been something which looked like *Solanum* 'hierba mora'."

"Jan. 1, 1954.—In the evening, just prior to a fête-champêtre at Centinela where we lived in an agricultural college, about 5 centiliters of the infusion were taken. After approximately three hours, I noticed extreme dryness in my mouth; a strong urge to urinate was also felt. This was unfortunately impossible as the urine emerged just one drop at a time though repeated attempts were made."

Somewhat later, possibly three or more hours after the consuming of the infusion, I felt a 'heavy' intoxication. This was not like alcoholic inebriation, where you have rather happy and agreeable thoughts and events before a hazy state occurs. This was an immediate and almost complete loss of memory."

"Without my knowledge, my friend, Dr. Earl E. Smith, Beltsville, brought me home and put me to bed. In my journal I noted that I 'missed my chance with the girl', but later Smith consoled me and told me that the girl in question had a face like a horse although she had a beautiful body. Even this mistake might have been due to the infusion."

Jan. 3, 1954.—I awoke in the middle of the night with, as I remember, a kind of claustrophobic feeling, or it might have been a normal feeling that woke me up followed by the claustrophobia. I could not find the toilet and started to run down the passage. According to what I remember, it was dark and I hit walls and doors. My nose started to bleed. In one way or another, I went into Smith's room and when I felt something that seemed to be a bed, I crept down—to Smith's surprise and alarm. He later told me that with force he managed to lead me back to my own room and locked me in."

"I had quite a hangover the next morning, although I was fairly clear in the head, and wanted to take part in a previously planned excusion. At breakfast . . . I suddenly spoke to those present in a completely unknown language without looking at anyone present. Unfortunately, I remember nothing from this conversation, nor with whom I thought I spoke. Possibly it was one of the last of the Araucarian hechiceros who finally had found a chap to talk with. What I remember is that I suddenly jumped out of my chair, thinking that someone wanted to beat me. It was my own hand which hung on the back of the chair which frightened me. I was unable to read a message on a piece of paper I had received before breakfast."

"During the day-long excusion, I was periodically fairly clear, periodically drowsy. During these latter periods, I saw the forest around the road as some kind of Russian boyar-ballet in heavy costumes. Oddly enough, green was the dominant colour."

"In the afternoon I felt fairly well restored, though my working capacity was less than usual. I still could not read."

"*Jan. 4, 1954.*—Smith and I proceeded on our trip at a normal pace, though in the morning I still could not read. In the evening, I wrote notes in my journal, but I had difficulties in keeping to the lines."

"*Jan. 5, 1954.*—Completely normal sight. Normal conditions."

It has only recently been confirmed that *Latua* is employed by the *machi* (medicine men) as a psychoactive agent. This was revealed to Sr. Rolando Toro, a psychologist from Santiago, who attended a *machitun* in Chiloé, in which *Latua* played an integral part of the ceremony. His account follows:

"*Latua* is used in an infusion by the shamans or curanderos, who ingest it during nocturnal ceremonies of a magical nature. After drinking the infusion at 20 to 30 minute intervals, they slowly begin to sing and dance in a circle. The chants are variations on the word latué:"

"Latué—latué—la—tué"
"La—la—la tué"
"Tué"
"La—tué"
"La—a—a—a" (slowly)
"La—tué—la—tué—la—tué (fast).

"Their movements are monotonous and consist in marking the rhythm by stomping their feet on the ground, along with movements of the head with the arms hanging like wings. The movements are not graceful but rather rigid, like those of catatonia. The dances last for four to six hours with intermittent prayers:"

"Con un tizón ardiendo"
"Cristo quema el mal"
"de vientro de N."

"With a firebrand
"Christ burns the evil
"from the belly of N." (here the name of the patient).

"The cure consists in driving the demons from the body of the patient. To do this, he is slapped with branches of *palqui* (*Cestrum Parqui* L'Her., Solanaceae) and is made to drink a potion which makes him vomit. Then his face is covered with the genital skin of a goat. The cure embraces every type of physical and mental infirmity and is always given at night. These meetings are equivalent to a witches' sabbath with curative ends."

We do not know how widespread is this use of *Latua*,

although it must be known to most of the *machis* in the region in view of their familiarity with the medicinal flora. It is of interest to note the Christian influence which has been incorporated into a primarily indigenous ceremony. This mixture of religious elements is reminiscent of shamanistic practices in other areas, such as the ceremonial use of *ayahuasca* and *San Pedro* in Peru, magic mushrooms and *peyote* in Mexico. *Latua* must now be appended to the growing list of plants used in magico-religious rites for hallucinogenic purposes.

From Timothy Plowman, Lars Olof Gyllenhaal and Jan Erik Lindgren "*Latua Pubiflora* Magic Plant From Southern Chile." *Botanical Museum Leaflets.* Harvard University. Cambridge, Massachusetts. (Nov. 12, 1971) vol. 23 #2, pp. 62, 70, 75-77, 81-82. (Reprinted by permission) [The co-authors are responsible only for the chemical part of the article.]

Tabaco

GONZALO FERNANDEZ de OVIEDO y VALDEZ

Of *Tabaco* or the Smoking Habit of the Indians On the Island of Hispaniola; and of the Kind of Bed in which They Sleep.

The Indians of this island, among others had a very wicked vice, to wit: they took smoke which they call *tabaco,* in order to go out of their senses. They did this with the smoke of a certain herb which, as I have been able to ascertain, is of the quality of henbane, but in form and shape it looks different to the eye; for this herb is four or five palms in height, and its leaves are broad, thick, soft and downy. Its green somewhat resembles that of the leaves of the oxtongue or *Bugloss,* so called by herbalists and doctors. The plant, as I say, is somehow similar to henbane. They take it in this manner: the caciques, or principal men, had some hollow sticks about a span long, and as thick as the small finger on a hand. These pipes had two hollow tubes, merging into one . . . and the whole forms one piece. And these two [ends] they put in their nostrils and the other [end] in the smoke of the burning herb. This instrument was smooth and well polished. They burned the leaves of this herb, folded up or wrapped in the way in which court pages are used to blow out their smokes, and they inhaled the smoke once, twice, thrice, or as often more as they could endure it, until they lost their senses and for a large space of time they lay stretched out on the ground, drunk and in a deep and profound sleep. And the Indians who could not get these little sticks took this smoke through a *calamus* or through little canes of the common reed grass. It is to this instrument with which they take the smoke the Indians give the name *tabaco* and not to the herb or the profound stupor in which they fall, as some have believed. The Indians considered this herb very precious and grew it in their gardens and farmlands for the aforementioned purpose. They let it be understood that taking this herb and its smoke was not only a healthy, but also a very holy thing to do. And as the cacique or chief lies on the ground, his wives (who are many) take him and put him on his bed or hammock, if he has ordered them to do so before he fell down. But if he has not told them or disposed so before, he wishes to be left on the ground until the drunkenness and sleepiness passes off. . . .

From Captain Gonzalo Fernandez de Oviedo y Valdés (1478-1552) *Historia General Y Natural de las Indias, Islas y Terra-Firme del Mar Oceano* . . . (Translation: General and Natural History of the Indies, Islands and Continent of the Ocean.) (In three Parts) 4 volumes. Imprenta de la Real Academia de la Historia, Madrid. (1851-55) Part 1, book V, ch. ii.) (First edition: Toledo (1526) and Sevilla (1535)

Petun *(Nicotiana* spp.)

ANDRÉ THEVET

. . . There is another peculiar herb, called *petun* by the Indians in their language. They most commonly carry this herb about them, for it is marvelously useful for many things. It is like our Bugloss. The Indians pick this herb very carefully and let it dry in the shade within their small huts. They use it in this manner: when dried they wrap a quantity of this herb in a very large leaf of a palm tree, and thus they make rolls the length of a candle; they set fire to one end [of these rolls] and receive the smoke thereof by their nose and by their mouth. The Indians say that this smoke is very wholesome for clearing and consuming superfluous humors of the brain. Moreover, if taken in this manner it satisfies hunger and thirst for some time. They also commonly use it if they have to discuss some matters among themselves: they draw in the smoke and then speak. . . . It is true that if too much of this smoke or perfume is taken, it rises to their heads and harms them in the same manner as strong flavored wine does. Christians who now live there, became extremely fond of this herb or perfume: although, before being accustomed to it, this herb may not be taken without harm, for its smoke causes perspiration, and weakness and even makes the users fall into syncopes,. and this I personally experienced. This is not as strange as it seems, for there are many other fruits offending the human brain, although, if eaten, they have a good and pleasant taste. . . .

From André Thevet (1501-1590) *Les Singularitez de la France Antartique, autrement nommée Amérique* . . . (Translation: Singularities of antarctic France, otherwise called America.) Christophle Plantin. Antwerp. (1558) ch. xxxii. (First edition: Paris. (1557)

Tabacco *(Nicotiana* spp.)

GIROLAMO BENZONI

. . . In this island [Hispaniola], as also in other provinces of these new countries, there are some bushes, not very large, like reeds, that produce a leaf in shape like that of the walnut, though rather larger, which (where it is used) is held in great esteem by the natives, and very much prized by the slaves whom the Spaniards have brought from Ethiopia.

When these leaves are in season, they pick them, tie them up in bundles, and suspend them near their fire-place till they are very dry, and when they wish to use them, they take a leaf of their grain (maize) and putting one of the others into it, they roll them round tight together; then they set fire to one end, and putting the other end into the mouth, they draw their breath up through it, wherefore the smoke goes into the mouth, throat, the head, and they retain it as long as they can, for they find a pleasure in it, and so much do they fill themselves with this cruel smoke, that they lose their reason. And there are some who take so much of it, that they fall down as if they were dead, and remain the greater part of the day or night stupefied. Some men are found who are content with imbibing only enough of this smoke to make them giddy, and no more. See what a pestiferous and wicked poison from the devil this must be. It has happened to me several times that, going through the provinces of *Guatemala* and *Nicaragua,* I have entered the house of an Indian who had taken this herb, which in the Mexican language is called *tabacco,* and immediately perceiving the sharp fetid smell of this truly diabolical and stinking smoke, I was obliged to go away in haste, and seek some other place.

In *La Española* and other islands, when their doctors wanted to cure a sick man, they went to the place where they were to administer the smoke, and when he was thoroughly intoxicated by it, the cure was mostly effected. On returning to his senses he told a thousand stories, of his having been at the council of the gods and other high visions. . . .

From Girolamo Benzoni (b. 1519) *Historia Del Mondo Nuovo.* (Translation: History of the New World). F. Rampazetto. Venice. (1565) leaves 54v to 55v. (English translation: London. (1857) Hakluyt Society. 1st Ser. #2) pp. 80-82. (Reprinted 1964).

Tabaco or Pecielt (*Nicotiana* sp.)

NICHOLAS MONARDES

Of the Tabaco, and of His Greate Vertues

THIS hearbe which commonly is called Tabaco, is an Hearbe of muche antiquitie, and knowen amongst the Indians, and in especially among theym of the newe Spaine [Mexico], and after that those countries were gotten by our Spaniardes, beyng taught of the Indians, they did profite themselves of those thinges, in the Woundes whiche they received in their Warres, healyng themselves therewith, to the greate benefite of them. . . .

The proper name of it amongest the Indians is Pecielt, for the name of Tabaco is geven to it of our Spaniardes, by reason of an Ilande that is named Tabaco. . . .

One of the merveilles of this Hearbe, and that whiche doeth bryng moste admiration, is the maner how the priestes of the Indias did use of it, whiche thei did in this maner: when there was amongest the Indians any maner of businesse, of greate importaunce, in the whiche the chiefe gentlemen called Casiques, or any of the principall people of the Countrie, havyng necessities to consulte with their Priestes, in any businesse of importaunce: then thei wente and propounded their matter to their chief Prieste, forthewith in their presence, he toke certain leaves of the Tabaco, and caste theim into the fire, and did receive the smoke of them at his mouthe, and at his nose with a Cane, and in takyng of it, he fell doune uppon the grounde, as a dedde manne, and remainyng so, accordyng to the quantitie of the smoke that he had taken, and when the hearbe had doen his woorke, he did revive and awake, and gave theim their aunsweres, accordyng to the visions, and illusions which he sawe, whiles he was rapte of the same maner, and he did interprete to them, as to hym semed beste, or as the Devill had consailed hym, givyng theim continually doubtfull aunsweres, in such sorte, that how soever it fell out, thei might saie that it was the same, whiche was declared, and the aunswere that thei made.

In like sorte the reste of the Indians for their pastyme, doe take the smoke of the Tabaco, for to make theim selves drunke withall, and to see the visions, and thinges that doe represent to them, wherein thei dooe delight: and other tymes thei tooke it to knowe their businesse, and successe, because conformable to that, whiche thei had seen beeying drunke therewith, even so thei might judge of their businesse. And as the Devill is a deceiver, and hath the knowledge of the vertue of Hearbes, he did shewe them the vertue of this Hearbe, that by the meanes thereof, thei might see their imaginations, and visions, that he hath represented to theim, and by that meanes doeth deceive them.

The Indians of our Occidentall Indias, doeth use of the Tabaco for to take awaie the wearines, and for to take lightsomnesse of their laboure, whiche in their Daunces thei bee so muche wearied, thei remaine so wearie, that thei can scarcely stirre: and because that thei maie laboure the nexte daie, and retourne to doe that foolishe exercise, thei dooe take at the mouthe and nose, the smoke of the Tabaco, and thei remaine as dedde people, and beeyng so, thei bee eased in suche sorte, that when thei bee awakened of their slepe, thei remaine without wearinesse, and maie retourne to their labour as muche more, and so thei dooe al waies, when thei have neede of it: for with that slepe, thei doe receive their strength, and bee muche the lustier.

The Indians dooeth use the Tabaco, for to suffer the drieth, and also for to suffer hunger, and to passe daies with out havyng nede to eate or drinke, when thei shall travaile by any deserte, or dispeopled Countrey, where thei shall finde neither water, nor meate. Thei doe use of these little baules, which thei make of this Tabaco, thei take the leaves of it, and doe chewe it, and as thei goe chewyng of them, thei goe mingling with them certaine pouder, made of the shelles of Cockles burned, and thei goe minglyng it in the mouth all together, untill that thei make it like to dowe, of the whiche thei make certaine little baules, little greater than Peason, and thei put theim to drie in the shadowe, and after thei keepe them, and use them . . .

From Nicholas Monardes, Physician of Seville (c. 1512-1588) *Joyfull Newes Out Of The Newe Founde Worlde.* Written in Spanish and Englished by John Frampton, Merchant Anno. (1577) . . . 2 volumes. Constable & Co. Ltd. London. Alfred A. Knopf, (1925) (The Tudor Translations. Second Series. Edited by Charles Whibley, IX) I, pp. 75f., 85-86 & 90.

Tobacco (*Nicotiana* sp.)

KING JAMES I

. . . And now good Countrey men let us (I pray you) consider, what honour or policie can moove us to imitate the barbarous and beastly maners of the wilde, godlesse, and slavish *Indians,* especially in so vile and stinking a custome? . . .

Why doe we not as well imitate them in walking naked as they doe? in preferring glasses, feathers, and such tpyes, to golde and precious stones, as they do? yea why do we not denie God and adore the Devill, as they doe? . . .

Have you not reason then to be ashamed, and to forbeare this filthie noveltie, so basely grounded, so foolishly received and so grossly mistaken in the right use thereof? In your abuse thereof sinning against God, harming yourselves both in persons and goods, and taking also thereby the markes and notes of vanities upon you: by the custome thereof making your selves to be wondered at by all forraine civil Nations, and by all strangers that come among you, to be scorned and contemned. A custome lothsome to the eye, hateful to the Nose, harmefull to the braine, dangerous to the Lungs, and the blacke stinking fume thereof, neerest resembling the horrible Stigian smoke of the pit that is bottomlesse.

From King James I (1566-1625) *A Counterblaste to Tobacco.* Imprinted by R[obert] B/arker/ London. (1604) fol. pp. Blv, B2r, D2r.

Tsángu (Jib.) Tabacu (Qu.) *(Nicotiana ondulata)*

RAFAEL KARSTEN

. . . The importance which the Jibaros ascribe to tobacco is enormous; from this plant they prepare their most powerful and most indispensable medicine. . . . The tobacco which the Jibaros cultivate is used by them only for strictly ceremonial purposes. On the other hand, the Indians would never use the tobacco bought from the whites for their religious ceremonies; they distrust it in the same way as they distrust other articles received from their hereditary enemies. The Indian tobacco, badly cultivated as it is, is not particularly strong; but since it is consumed in great quantities, its effects are still powerful enough.

The Jibaros usually take tobacco in a liquid form, but in some cases they smoke it, making big cigars of the leaves. In the former case the leaves are either boiled in water, or chewed in the mouth and carefully mixed with saliva. When used at the great feasts the medicine is always prepared with saliva, which is supposed to enhance its supernatural effects. This medicine is prepared by an old man who conducts the ceremonies, and he has also to dispense it to the persons in whose honour the feast is held. The men on these occasions always receive the tobacco juice through the nose, the women again through the mouth.

After two or three doses the effects of the medicine are generally noticeable. The Indian turns pale and the whole body begins to tremble. After repeated doses giddiness and fainting-fits frequently occur, especially in women. At last the person in question becomes quite drugged and falls into a long sleep filled with peculiar dreams.

Tobacco is taken on the following occasions: first, as a universal remedy against disease and evils of any kind; secondly, as a prophylactic means of enhancing the magical power of the body, particularly to resist the evil spirits; and lastly, as a real narcotic to produce dreams. In all three cases the plant has purely magical or religious significance.

. . . . . Just as the Indians take *natéma* and *maíkoa* "in order to have dreams", so they drink tobacco water for the same purpose.

[Tobacco is the particular medicine of sorcerers and medicine-men. It is indispensable at the great tobacco feasts, that of the men and that of the women, and finally at the great victory feast of the Jibaros.]

From Rafael Karsten (1879-1956) "The Head-Hunters of Western Amazonas. The Life and Culture of the Jibaro Indians of Eastern Ecuador and Peru." Helsingfors. Societas Scientiarum Fennica. *Commentationes Humanarum Litterarum* VII (1935) #1 pp. 441-43.

Tobacco (*Nicotiana* spp.)

JOHANNES WILBERT

Tobacco may be one of several vehicles for ecstasy; it may be taken in combination with other plants . . . to induce narcotic trance states; or it may represent the sole psychoactive agent employed by shamans to transport themselves into the realm of the supernatural, as is the case among the Warao of the Orinoco Delta in Venezuela. That Warao shamans smoke enormous "cigars" as much as 50 to 75 centimeters long as been known since early contact times, but the meaning of tobacco in Warao intellectual culture has often gone unnoticed. . . .

The Warao believe they inhabit a saucer-shaped earth surrounded by a belt of water. The "stepped" celestial vault covers both earth and ocean and rests on a series of mountains situated at the cardinal and intercardinal points. Much of a Warao Indian's life is spent in propitiating a number of Supreme Spirits *(Kanobos)* who inhabit these mountains at the ends of the world and who require nourishment in the form of tobacco smoke from the people.

The priest-shaman *(wishiratu)* visits these spirits in his dreams or in a tobacco-induced trance and, on returning from such a visit, transmits the message of the Supreme Spirits to the community. One of the four major spirits is usually present among the people in form of a sacred stone. The annual *moriche* festival, called *nahanamu,* over which the priest-shaman presides, is celebrated in propitiation of the Supreme Spirits who request that the ceremony be held and who will protect the community if their command is heeded. Sickness is believed to be caused by one or another of the deified *Kanobos,* who thereby expresses his dissatifaction with man and sends his *hebu* (spirit) to do harm or even kill. Children especially are subject to such attacks. The priest-shaman is the only one who can intervene as curer because only he can relate directly to the Supreme *Kanobos.*

In addition to the priest-shaman there are two other important religious practitioners among the Warao. The "light" shaman is known as *bahanarotu.* He presides over an ancient cult of fertility called *habisanuka.* . . . the *bahanarotu* travels in his dream or tobacco-induced state to an eastern part of the cosmic vault. The celestial bridge of tobacco smoke which he frequents and maintains between his community and the eastern Supreme *Bahana* (spirit) guarantees abundance of life on earth. In their aggressive shamanic role, *bahanarotus* spread sickness and death among their enemies by hurling magic arrows at them. Only a friendly *bahanarotu* can assuage such misfortune, through the use of tobacco and the widespread traditional shamanic technique of sucking out the illness-causing foreign bodies magically introduced by the malevolent sorcerer.

The "dark" shaman, known as *hoarotu,* maintains the connection between the Warao in the center of the universe and the powers of the West. This connection became severed in ancient times and can be re-established only by the *hoarotu.* The spirit beings in the West subsist through their medium, the dark shaman, on the blood and flesh of man. To procure this human food for his masters, the *hoarotu* kills his victims by means of magic projectiles, again through the medium of tobacco smoke. . . .

The "history of consciousness" of the Warao as a people has its origin in the "House of Tobacco Smoke", created *ex nihilo* by the Creator Bird of the Dawn. The House of Smoke is the birthplace of "light" shamanism, called *bahana.* Its materialization by means of solidified tobacco smoke took place through the conscious act of a bird spirit, who at the beginning of time arose as a young man in the East. The radiant body of this youth, his weapons, and his shamanic rattle were all made of tobacco smoke. . . .

Since *bahanarotus* can see *bahanas* in the dark, they sometimes get together for a tobacco seance to play the supernatural Game of *Bahana,* before the eyes of the awestruck villagers. Exhaling puffs of smoke, they send the four pieces of the *Bahana* Game [crystal, hair ball, rocks, and tobacco smoke] one after the other to travel like luminous bodies through the dark house. The quartet of bahana *spirits* [the Black Bee, Wasp, Termite, and Honey Bee] spirits delight in this game. Generally they are said to be the aforementioned "power objects"—rock crystals, hair, rocks and puffs of tobacco smoke—but a bullet, a piece of glass, or a button will also serve as a magic projectile. They drift through the air, seeking out one or the other of the spectators, but since this is only a game, they do not enter his body. The people in the room are fearful of this supernatural demonstration of shamanic power. But the "fathers" of the roving *bahanas* always call their "sons" back if the game threatens to get out of hand. They blow tobacco smoke to intercept the flight of the spirits and put them back in their basket.

Bahanarotus travel frequently to the House of Smoke in the East, and when they die they go to live there forever.

The Scarlet Macaw *(Ara chloroptera)* is the Supreme *Hoa* spirit who rules over the Abode of Darkness, called *Hoebo*. This place is situated at the end of the world to the West. Here live all the souls of deceased "dark" shamans, the *hoarotus,* as being half human and half animal. The stench of human cadavers and clotted blood saturates the air, and the stream of *hoaratu* shamans who come from all parts of Waraoland with cadavers hanging head down from their shoulders is endless. It has to be endless if the Supreme *Hoa* and spirit companions, called *hoarao,* are to continue living: the former by eating human hearts and livers, the latter by devouring the bodies. All *hoaroa,* in the *Hoebo* drink human blood from a gigantic canoe made of human bone.

The abode of Darkness has existed since the beginning of time. . . .

. . . [Some *hoarotus* kill, but a friendly *hoarotu* can prevent death.] . . . In some communities there is nearly perpetual competition between *hoarotus* that kill and others that cure—lucky the village that can rely on a powerful *hoarotu* who knows how to keep the Scarlet Macaw and his spirits of the West appeased with a minimum of sacrifices, while maintaining the strength of his own group by saving his fellows from the *hoa* snare [i.e. the tobacco smoke which enters the chest of a victim] of malevolent *hoarotus.*

Throughout their lifetime *hoarotus* travel often to the *Hoebe* in the West, always using tobacco as their means of ecstasy. They too have a house in the Otherworld in which they will dwell forever after death. But while the house of the *bahanarotu* is in the East, the land of light, that of the *hoarotu* is in the West, the realm of darkness.

From Johannes Wilbert (Professor of Anthropology and Director of the Latin American Center, University of California at Los Angeles) Article entitled "Tobacco and Shamanistic Ecstasy among the Warao Indians of Venezuela", from *Flesh of the Gods,* edited by Peter T. Furst. (c) 1972 by Praeger Publishers, Inc., New York. pp. 57, 58, 59, 60, 65, 72, 73 & 78. (Reprinted by permission)

Narcotics of Uncertain Origin

Carapullo [*Carapucho?*]

AMÉDÉE FRANÇOIS FREZIER

I should not forget to mention here some details about certain plants which were named to me by some trustworthy persons. In Peru grows an herb called *carapullo* [*carapucho?*] which has a stem like a grass and produces an ear; a decoction of the latter, when drunk, causes a delirium lasting for several days. This is used by the Indians to find out the natural disposition of their children. While the children are under the effect of this decoction, the Indians place next to them the tools of all kinds of profession that they might take up. For example, next to their daughters, they place distaffs, wools, scissors, linen, kitchen utensils, etc.; next to their sons, horse harnesses, awls, hammers, etc. The tools to which the children most attach themselves during the delirium will be the sure sign of the profession for which they will be fitted; thus a French surgeon assured me, who had witnessed this rare thing.

From Amédée François Frezier (1682-1773) *Relation du Voyage de la Mer du Sud aux Côtes du Chilly et du Perou, fait pendant les Années 1712, 1713, & 1714.* J. G. Nyon, etc. Paris. (1716) p. 213.

JEAN BAPTISTE LABAT

"An Extraordinary Great Number of Cases of Madness on Martinique in 1699."

I do not know under which vicious star Martinique has been this year, but never before such disorder, nor such a number of mad persons has been seen. Many people, without any fever or any other visible illness, became delirious and bereft of reason and began to run through the streets doing a thousand of follies. . . .

Some drowned themselves, while others broke their necks falling from trees or cliffs which they had climbed in order to practice flying in the air. Prison and applying the cudgel brought some of them back to reason.

From Jean Baptiste Labat (1663-1738) *Nouveau Voyage aux Isles de l'Amérique . . .* 6 vols. P. F. Giffart. Paris. (1722) IV, ch. xii.

Curupa

(Formerly believed to be a *Piptadena* species of the Pulse Family.)

CHARLES MARIE DE LA CONDAMINE

(July 1743) . . . The Omagua Indians make much use of two kinds of plants. One of these plants is called *floripondio* by the Spaniards. The flower of this resembles a reversed bell, as Father Feuillée described it. The other plant is called *curupa* by the Omaguas in their language. Both plants have purging properties. However, these Indians use these plants to get intoxicated for twenty four hours, during which time they have the strangest visions. The *curupa* is also taken ground up to a powder, in the way as we are taking tobacco, however with more apparatus. They use a reed pipe of "Y" shape, ending in a fork, and the ends of this fork they put into each of their nostrils. This operation is followed by violently drawing in their breath which causes them to make grimaces, most ridiculous to the eyes of a European who wants everything to be conform to his customs.

From Charles Marie de La Condamine (1701-1774) *Relation abrégée d'un Voyage fait dans l'Intérieur de l'Amérique Meridionale . . .* (Translation: An abridged account of a journey made to the interior of South America.) Veuve Pissot. Paris. (1745) p. 43.

JOSEPH SKINNER

. . . One of the principal particulars which refer to natural history is the disease, or furious madness, as it is termed by

Dr. Cosme Bueno (1711-1798), which attacks both men and beasts, in the town of Tatasi, belonging to the department of Chichas. On the first access of this frenzy, there are not sufficient powers to restrain the unfortunate victim who, forgetful of all shame and human necessity, forsakes his bed, flees from the habitations of men, runs impetuously over the mountains in the environs, and rushing from precipice to precipice, at length hurls himself from the summit of the steep rock. It usually happens that, in falling from a considerable height, he is bruised to death; but if, by rare casualty, he survives, in proportion as he recovers his bodily health, the mental powers return to their just equilibrium, and there is no longer any vestige of this terrible malady.—Experience has demonstrated, that the animals originally brought from Europe, such as horses, oxen, sheep, &c. are the only ones affected by this malady, to which those that are natural to the country, such as vicuñas, huanacos, &c. are not liable.

From [Joseph Skinner] *The Present State of Peru.* Printed for R. Phillips. London. (1805) pp. 332-33.

MARTIN DOBRIZHOFFER

"Of A Certain Disease Peculiar To The Abipones."

During an eighteen years' acquaintance with Paraguay and its inhabitants, I discovered a disease amongst the Abipones Nakaiketergehes, entirely unknown elsewhere. This disease affects the mind more than the body, . . . They sometimes begin to rave and storm like madmen. The credulous and superstitious crowd think them reduced to this state by the magic arts of jugglers [here women sorcerers], and call them *Laopar̃aika.* These persons . . . betray their madness chiefly at sun-set. The distracted persons suddenly leap out of their tents, run into the country on foot, and direct their course straight to the burying-place of their family. In speed they equal ostriches, and those who pursue them on the swiftest horses can hardly overtake and bring them home. Seized with fury in the night, they burn with the desire of committing slaughter somewhere; and for this purpose snatch up any arms they can lay hold of. . . . The hordesmen, as they can neither calm the furious man, nor keep him at home, suffer him to go out into the street, armed with a stick, and accompanied with as many people as possible. . . .

The insane person strikes the roof and mats of every tent again and again with the stick, none of the inmates daring to utter a word. If through the negligence of his guards, or his own cunning, he gets possession of arms, Heavens! what a universal terror is excited! . . .

Persons seized with this madness take scarcely any food or sleep, and walk up and down pale with fasting and melancholy: you would imagine that they were contemplating some new system of the figure of the earth, or studying how to square the circle. By day, however, they betray no signs of alienation of mind, nor are they to be feared before evening. A person of this description, who was very turbulent at night, visited me in the middle of the day. In familiar conversation I asked him who it was that disturbed the rest by his furiousness every night. He replied with a calm countenance, that he did not know. . . . Sometimes many persons of both sexes began to rave at once; sometimes one, and often no one was in this deplorable state. This madness lasted eight, fourteen, or more days, before tranquility and intellect were restored. All the Abipones subject to this malady, whom I have known, were uniformly of a melancholy turn of mind, always in a state of perturbation from their hypochondriac or choleric temper, and of fierce, threatening countenance. When this bile was excited by bad air, or immoderate drinking, it is neither strange nor surprising, that derangement and raving madness ensued. The stupid or ignorant alone attribute that to magic art, which is solely to be ascribed to the fault or strength of nature.

We have found the fear of death a powerful antidote to the licence of raving amongst the Abipones. Within a few days the number of mad persons increased unusually: one of them in the dead of the night got through the fence, and was stealing into our house, but was carried away by people who came to our assistance. Alaykin, the chief Cacique, being informed of our danger, called all the people into the market-place next day, and declared, that if any one henceforward took to raving, he should immediately put to death all the female jugglers, as well as the insane themselves. From that time I never heard of any more tumults occasioned by these furious persons. . . . I never can believe with the savages, that a magical charm was the cause of their insanity.

From Martin Debrizhoffer (1717-1791) *An Account of the Abipones, an Equestrian People of Paraguay* [Translated] From the Latin [by Mrs. Sara Henry Colleridge] In Three Volumes. John Murrey. London. (1822) II, ch. xxii, pp. 233-37.

INDEX OF LATIN NAMES OF GENERA AND SPECIES OF PLANTS

INDEX OF VERNACULAR NAMES OF PLANTS AND PLANT PRODUCTS